D1671450

The Enhancement of Gear Quality through the Abrasive Flow Finishing Process

The Enhancement of Gear Quality through the Abrasive Flow Finishing Process

By

Anand Petare, Neelesh Kumar Jain and I.A. Palani

Cambridge Scholars Publishing

The Enhancement of Gear Quality through the Abrasive Flow Finishing Process

By Anand Petare, Neelesh Kumar Jain and I.A. Palani

This book first published 2023

Cambridge Scholars Publishing

Lady Stephenson Library, Newcastle upon Tyne, NE6 2PA, UK

British Library Cataloguing in Publication Data
A catalogue record for this book is available from the British Library

ISBN (10): 1-5275-9443-2
ISBN (13): 978-1-5275-9443-2

Table of Contents

PREFACE

A gear is a modified form of a wheel. Globally more than 10 billion gears of different types are bought per year. This makes the manufacturing and finishing of gears very important in terms of technological, environmental, and economical aspects. Continuous research and innovations are required to develop technically superior, environment-friendly, productive, and affordable processes for the manufacturing and finishing of different types of gears made of wide-ranging materials. The abrasive flow finishing (AFF) process is such an advanced finishing process as it fulfills most of the requirements of a modern gear finishing process, but its potential for gear finishing has not yet been fully exploited.

The objective of this book is to provide details of the extensive work done to the AFF process for the high quality finishing of cylindrical (i.e. spur) gears and conical (i.e. straight bevel) gears. It consists of ten chapters. The first chapter introduces different aspects of the abrasive flow finishing (AFF) process, including its types, different parameters, advantages, limitations, and some typical applications. Chapter 2 introduces gears, their materials, the manufacturing process and quality evaluation parameters, traditional gear finishing processes, and a summary of their limitations. Chapter 3 presents a comprehensive review of past work on the finishing of different types of gears by the AFF process, the use of laser texturing, the effect of gear finishing on noise and vibration reduction, the existing gaps, and the objectives of the present work and the methodology adopted to meet them. Chapter 4 describes the development of the AFF machine for the AFF process, the fixtures for finishing and holding spur and straight bevel gears, and the details of the selected materials for finishing the spur and straight bevel gears as well as their specifications. Chapter 5 details the design, planning, and conduct of experimental investigations in different stages along with an evaluation and characterization of the surface roughness, microgeometry, surface morphology, microhardness, and wear indicating the parameters of spur and straight bevel gears finished by the AFF process. Chapter 6 presents the obtained results, their analyses, and conclusions from different stages of the experiments. Chapter 7 presents details of the optimization of the AFF process parameters and experimental validations of the optimization results. Chapter 8 describes the laser texturing of spur and straight bevel gears and a comparative study of AFF of the untextured and laser-textured gears in terms of microgeometry error, surface roughness, surface morphology, wear characteristics, and microhardness. Chapter 9 presents an evaluation and analysis of the functional performance parameters and noise and vibration characteristics of the gears finished by the AFF along with their comparison with unfinished gears. Chapter 10 concludes this book by presenting significant achievements and conclusions, and directions for future research.

It is expected that this book will be of great help to the manufacturers and users of different types of gears who are looking for a better gear finishing process.

Anand Petare
Neelesh Kumar Jain
I A Palani

NOMENCLATURE

C_{av}	Volumetric concentration of the abrasive particles in the AFF medium (%)
d_a	Diameter of the abrasive particles (μm)
fi	Tooth-to-tooth composite error (μm)
ff_a	Profile form error (μm)
ff_β	Lead form error (μm)
fh_a	Profile angle error (μm)
fh_β	Lead angle error (μm)
F	Applied normal load (N)
F_a	Total profile error (μm)
F_β	Total lead error (μm)
F_i	Total composite error (μm)
F_p	Total pitch error (μm)
F_r	Radial runout determined by the microgeometry error measurement (μm)
F_{rf}	Radial runout determined by the double flank roll testing (μm)
k_i	Specific wear rate (mm^3/Nm)
M_a	Abrasive particles size (mesh) $\left[M_a = \frac{15.24}{d_a\ (in\ mm)}\right]$
ΔN	Change in noise level (dBA)
O_c	Volumetric concentration of silicone oil in the AFF medium (%)
P	Extrusion pressure in the AFF process (MPa)
PRF_a	Percentage reduction in total profile error (%)
PRF_β	Percentage reduction in total lead error (%)
PRf_p	Percentage reduction in single pitch error (%)
PRf_u	Percentage reduction in adjacent pitch error (%)
PRF_p	Percentage reduction in total pitch error (%)
PRF_r	Percentage reduction in radial runout (%)
PRR_a	Percentage reduction in average surface roughness (%)
PRR_{max}	Percentage reduction in maximum surface roughness (%)
R_a	Average surface roughness value (μm)
ΔR_a	Change in average surface roughness value (μm)
R_{max}	Maximum surface roughness value (μm)

ΔR_{max}	Change in maximum surface roughness value (μm)
S	Total sliding distance in the wear test (m)
t	Finishing time in the AFF process (Minutes)
ΔV	Change in vibration level (mm/sec^2)
η	Viscosity of the AFF medium (kPa.s)
m_i	Mass loss during the wear test (mg)
ρ	Density of the gear material (kg/mm^3)

ACRONYMS

AFF	Abrasive Flow Finishing
AGMA	American Gear Manufacturers Association
ANOVA	Analysis of Variance
BBD	Box-Behnken Design
CCD	Central Composite Design
CNC	Computer Numeral Control
COF	Coefficient of Sliding Friction
DFA	Desirability Function Analysis
DIN	Deutsches Institut für Normung
DOE	Design of Experiments
DOF	Degree of Freedom
HRB	Rockwell Hardness at B Scale
HV	Vickers Hardness Number
LF	Left Flank
LTSBG	Laser Textured Straight Bevel Gear
LTSG	Laser Textured Spur Gear
MRR	Material Removal Rate
RF	Right Flank
RSM	Response Surface Methodology
SBG	Straight Bevel Gear
SG	Spur Gear
SEM	Scanning Electron Microscope
UAAFF	Ultrasonic Assisted Abrasive Flow Finishing
USG	Untextured Spur Gear
USBG	Untextured Straight Bevel Gear
WSEM	Wire Spark Erosion Machining

CHAPTER ONE

OVERVIEW OF ABRASIVE FLOW FINISHING

Abrasive flow finishing (AFF) is an advanced nano-finishing process developed by Extrude Hone Corporation (USA) in 1960. It is meant for finishing inaccessible and restricted areas in components, the simultaneous finishing of a large number of holes, the radiusing and deburring of complicated components, and the removal of the recast layer. Figure 1.1 depicts the working principle of the AFF process schematically. It uses hydraulically or mechanically controlled movements of a semi-solid, self-deformable, and pliable finishing medium in the form of a putty through the workpiece itself or a flow path formed between the workpiece and its holding fixture (Rhodes, 1991).

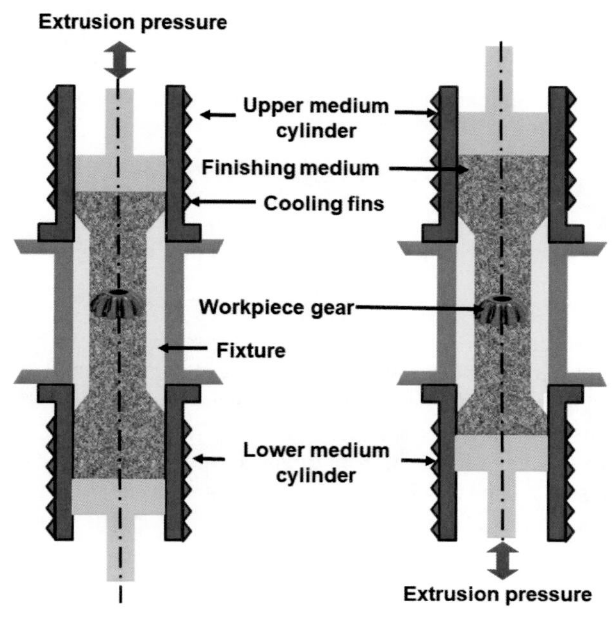

Fig. 1.1: Schematic of working principle of the abrasive flow finishing (AFF) process

A typical machine for the AFF process consists of four main subsystems, namely the (i) power supply unit to provide the required extrusion pressure hydraulically, pneumatically, or mechanically for the controlled movement of the AFF medium, (ii) supporting structure, (iii) workpiece fixture, and (iv) AFF medium. The power supply unit consists of hydraulic or pneumatic cylinders, cylinders containing the AFF medium, a power supply unit along with the control panel, direction control valves, a pressure regulator to set the desired pressure, and a pressure gauge. The supporting structure comprises plates and rods for supporting the hydraulic or pneumatic cylinders, cylinders containing the AFF medium, workpiece fixture, limit switches to set the stroke length, and a stroke counter. The workpiece fixture is designed according to the geometry and size of the component and the area to be finished. It can also be designed to finish the multiple components in a single setting of the AFF machine. It serves two functions: (a) to hold the workpiece tightly between the AFF medium-containing cylinders so it does not move while being finished at high extrusion pressure, and (b) to direct the flow of the AFF medium to those parts of the workpiece where finishing is required and restrict its flow where finishing is not required. Materials such as hardened steel, Teflon, Nylon, and Metlon can be used to make the workpiece fixture. The finishing medium used in the AFF process generally contains a uniform mixture of viscoelastic polymer, abrasive particles whose type and size depend on the type and hardness of the workpiece material and its finishing requirements, and the blending oil, which is used to maintain the desired viscosity of the AFF medium.

1.1 Types of AFF Process

The AFF process can be classified into the following four types according to the flow of the finishing medium and arrangements required for it:

- **One-way AFF:** In the one-way AFF process, the AFF medium is extruded from one end of the workpiece and collected at other end after one finishing cycle and has to be reinserted again for the next finishing cycle, as shown in the Fig. 1.2a. Although the process takes a very long time to finish a component and causes more wastage of the finishing medium, it maintains its freshness and thus its chemical composition and rheological properties do not change.

- **Two-way AFF:** The AFF medium in the two-way AFF process is moved back and forth continuously between two medium-containing cylinders, with the workpiece and its fixture sandwiched between them, as depicted in the Fig. 1.2b. The machine is heavier and costlier, and design of the fixtures for this process is also complicated. The AFF medium remains contained within the medium-containing cylinders until there is a change in the chemical composition and rheological properties of the AFF medium. There is a problem with leakage of the AFF medium.
- **Multi-way AFF:** The AFF medium is moved back and forth between four medium-containing cylinders in the multi-flow AFF process, as illustrated in the Fig. 1.2c. Two fixtures containing two workpieces are placed parallel to each other and an arrangement is made to transfer the AFF medium in cross-direction. Its machine is the heaviest and most complicated because it requires special arrangements to sequence the hydraulic operations.
- **Orbital AFF**: The AFF medium is moved back and forth between two hydraulically operated cylinders containing the AFF medium through a displacer whose shape is complementary to the workpiece geometry, as shown in the Fig.1.2d. Orbital vibrations are provided to the workpiece in the transverse direction. This process requires manufacturing the displacer, and its machine can provide orbital vibrations of the desired frequency and amplitude.

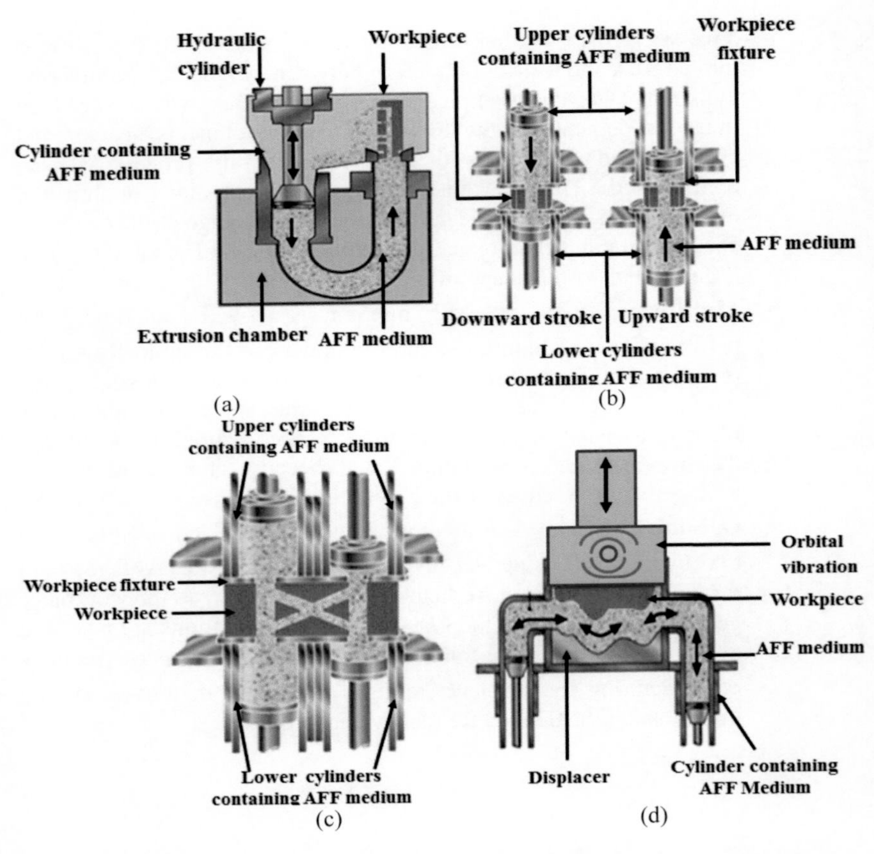

Fig. 1.2: Schematic views of four types of the AFF process: (a) one-way AFF; (b) two-way AFF; (c) multi-way AFF; and (d) orbital AFF (Petare and Jain, 2018) Reprinted with permission from Springer © 2018.

1.2 Parameters of the AFF Process

Required surface finish and material removal rate can be achieved by selecting a proper parametric combination of the AFF processes. They can be classified into three groups, namely, parameters associated with the AFF medium, AFF machine, and the workpiece (Jain, 2009).

1.2.1 Parameters Associated with the AFF Medium

Parameters associated with the AFF medium that affect the performance of the AFF process include type, size, and concentration of the abrasive particles, type of visco-elastic polymer, type and amount of the blending oil, and viscosity, temperature, and functional life of the AFF medium.

- **Type, size, and concentration of the abrasive particles:** The selection of the size and type of the abrasive used in the AFF medium depends on the type and finishing requirements of the workpiece material. Commonly used abrasives include alumina (Al_2O_3), silicon carbide (SiC), cubic boron carbide (CBN), diamond, and a mixture of these abrasives. Alumina is generally used for the softer workpiece materials and other abrasives for the harder workpiece materials. Fine abrasive particles are mostly preferred when a high-quality surface finish is required for the components whereas coarse abrasive particles are generally used for deburring and radiusing applications. An increase in MRR and surface roughness has been observed with an increase in abrasive particle size because higher abrasive particle sizes cause more indentation depth in the workpiece (Williams and Rajurkar, 1992). The volumetric concentration of abrasive particles is their % volume in the total volume of the AFF medium. The ratio of volume of abrasive particles to volume of viscoelastic polymer (or silly putty) can vary from 1:4 to 4:1; however, it has been suggested that a 1:1 ratio achieves a better surface finish (Perry, 1985). A higher concentration of abrasive particles increases the chance of jamming the cylinders containing the AFF medium. An optimum value of abrasive concentration exists because surface roughness decreases and MRR increases with an increase in abrasive concentration due to more abrasive particles being available for finishing the workpiece surface. However, after the optimum value, this trend reverses because the AFF medium tends to become rigid, thus losing its self-deformable nature (Sankar et al., 2009).
- **Type and concentration of the blending oil:** Blending oil is used to ensure the proper mixing of the visco-elastic polymer with the abrasive particles to form a bond between them and impart the

required viscosity to the AFF medium. Silicone oil, turpentine oil, and hydrocarbon-based oils are commonly used. An excessive amount of blending oil makes the AFF medium behave like a liquid and reduces its finishing capability, whereas too little imparts a high rigidity to the AFF medium, making it behave like a solid, which increases the chances of choking the medium-containing cylinders.

- **Viscosity of AFF medium:** The viscosity of the AFF medium depends on the type of viscoelastic polymer and the amount of blending oil used. A higher viscosity of the AFF medium increases the bonding strength between the abrasive particles and the polymer, causing more material removal from the workpiece and producing a uniform surface. A lower viscosity of the AFF medium reduces the bonding between the polymer and abrasive particles, which causes a rotation of the abrasive particles on the workpiece surface rather than sliding thus reducing MRR and increasing surface roughness. It is used for radiusing and deburring applications. For a given volume of AFF medium, viscosity increases with an increase in the concentration of the abrasive particles (Rajeshwar et al., 1994).
- **Temperature of the AFF medium**: Past research has revealed that the temperature of the AFF medium has a significant influence on its rheological properties (i.e. viscosity, viscoelasticity, thixotropy, etc.) and that it can increase by 30–70°C (Agrawal et al., 2005). The viscosity of the AFF medium decreases with an increase in its temperature due to a reduction in the bonding between the abrasive particles and viscoelastic polymer. This increases the settling tendency of the abrasive particles at the bottom of the medium-containing cylinder. Therefore, the temperature of the AFF medium should not be allowed to go beyond 100°C (Hull et al., 1992).
- **Preparation method of the AFF medium:** The AFF medium can be prepared by different methods such as kneading, rotary mixing, extrusion, and in the AFF machine itself without using the workpiece. The selected method should ensure a uniform mixing of the constituents of the AFF medium.
- **Functional life of the AFF medium:** The AFF medium is a consumable and its lifetime is typically about 250 hours, but the

actual value depends on the shape, size, and amount of finishing to be done (Przyklenk, 1986).

1.2.2 Parameters Associated with the AFF Machine

Parameters associated with the AFF machine include extrusion pressure, volume of the AFF medium (which is determined by the diameter and stroke length of the medium-containing cylinders), flow rate of the AFF medium, and the number of cycles or finishing time.

- **Extrusion pressure:** It is the summation of the force per unit contact area used in overcoming the friction between the AFF medium-containing cylinder and its piston and the force imparted on the AFF medium per unit area. It can be provided by a hydraulic, pneumatic, or mechanical power unit. Literature reveals that its value varies from 0.7 to 20 MPa (Jain and Adsul, 2000). An increase in the extrusion pressure increases the material removal rate (MRR), compressive stress on the AFF medium, and decreases average surface roughness (R_a) value (Williams and Rajurkar, 1992).
- **Volume of the AFF medium:** It is the volume of the AFF medium filled in the medium-containing cylinder and is determined by the cylinder's diameter and stroke length. A larger stroke length produces a larger slug length, which increases the abrasive action and MRR (Cheema et al., 2012).
- **Flow rate of the AFF Medium:** It is the volume of the AFF medium flowing per unit time. It depends on the viscosity of the AFF medium and the extrusion pressure.
- **Number of cycles or finishing duration:** One cycle of the AFF process consists of one upward stroke from the reference position and downward stroke back to the reference position by the piston of the medium-containing cylinder. It is a decisive factor to achieve the required surface finish. Its value depends on the desired value of surface roughness of the workpiece. Surface roughness of the workpiece decreases asymptotically with an increase in the finishing duration.

1.2.3 Parameters Associated with the Workpiece

Important parameters associated with the workpiece include the type and mechanical properties (such as ductility and hardness) of the workpiece material, type of manufacturing process used to make the workpiece, and initial surface condition.

- **Type and mechanical properties of the workpiece material:** The AFF process can be used for finishing both metallic (both ferrous and non-ferrous) and non-metallic materials (i.e. ceramics, composites, additive manufactured products). The ductility and hardness of the workpiece material significantly affect the finishing results and performance of the AFF process. Past studies have shown that finishing softer and more ductile materials by the AFF gives more MRR but a poorer surface finish than finishing the harder materials. This is due to a deeper indentation by the abrasive particles in softer materials than in the harder materials (Rhodes, 1991).
- **Type of manufacturing process used to make the workpiece:** AFF can be used to nano-finish milled, turned, ground, and wire-spark-erosion-machined (WSEMed) components (Loveless et al., 1994). It can achieve a very good improvement in the surface finish with the removal of a very small amount of material from the workpiece, and the finishing results are reliable, uniform, repeatable, predictable, and accurate.
- **Initial surface condition:** The initial condition of the workpiece surface significantly affects the finishing results by the AFF process. More variations in surface roughness produce more MRR but less improvement in the surface finish because at the start of the AFF process, the abrasive particles come in contact with the higher surface peaks and flatten them. However, after some finishing cycles, the flattening of a greater number of surface peaks takes place simultaneously. Loveless et. al (1994) compared the mechanism of material removal in the finishing of WSEMed, turned, milled, and ground surfaces by the AFF process and concluded that the WSEMed surface yields the best results because it has surface peaks with less height, micro-cracks, and loosely attached white layers.

1.3 Advantages of the AFF Process

The following are the advantages of the AFF process:

- It gives average surface roughness values as small as 50 nm with a dimensional tolerance as low as ± 5 μm. The thickness of the material removed is in the range of 1 to 10 μm.
- Multiple holes can be finished simultaneously in a single setup of the AFF process in minimum time. A minimum limit on the hole size that can be finished /deburred properly is 0.22 mm in diameter, and the largest size that has been finished is around 1000 mm in diameter.
- Finishing multiple parts simultaneously results in uniformity, repeatability, and predictability. Automatic AFF machines can finish thousands of parts per day, and the labor costs are reduced significantly by reducing manual handwork.
- It offers flexibility for changes in material, geometry, and dimensions of the workpiece, type of the AFF medium and abrasives, and the AFF process parameters, thus enabling finishing a variety of parts on the same setup just by changing the fixtures for holding the workpiece. It gives a general-purpose finishing solution for the mass production.
- The AFF medium can be selected according to the finishing requirements and the economic aspects.

1.4 Limitations of the AFF Process

The following are the limitations of the AFF process:

- Unable to improve surface irregularities such as out-of-roundness, tapered holes, blind holes, deep scratches, large bumps, cavities, etc.
- The material removal rate in the AFF process is very small hence it is not applicable for mass-scale material removal purposes.
- Uncertainty about the distribution of the abrasive particles in the AFF medium and difficulty controlling the rheological properties of the AFF medium.
- The design of fixtures changes every time according to the geometry of the part, and the development of a suitable restriction

and tooling to obtain the desired finish on the selected areas is a major challenge.

1.5 Applications of the AFF Process

The AFF process finds wide applications for finishing the various components used in different fields. Some major applications are listed below:

- Finishing of different components used in the aerospace, automotive, diesel, and turbine engine industries, i.e. deburring of aircraft valve bodies and spools, fuel spray nozzles, fuel control bodies, airfoil surfaces of impellers, common rail pipe fuel system of a diesel engine, outer rotor surface of the cycloidal pump, etc.
- Finishing of the fluid handling components of the food processing systems and the components manufactured by additive layer manufacturing processes.
- Aerospace applications include the removal of the thermal recast layer in the cooling holes of blades and disks, deburring of the fuel spray nozzles, and polishing cast surfaces of blades, compressor wheels, and impellers.
- Radiusing and deburring air-cooling holes of turbine disk and holes of combustion liner.
- Materials ranging from soft aluminum to hard nickel alloys, ceramics, and carbides can be successfully finished by the AFF process.
- Surface integrity problems such as residual stresses, metallurgical transformations, plastic deformations, heat-affected zones, recrystallizations, tearing, and cracking can be reduced or avoided.

References

[1] Agrawal, A., V. K. Jain, and K. Muralidhar. 2005. "Experimental determination of viscosity of abrasive flow machining media." *International Journal of Manufacturing Technology and Management.* 7(2–4), 142–156. doi:0.1504/IJMTM.2005.006828

[2] Cheema, M. S., G. Venkatesh, A. Dvivedi, and A. K. Sharma. 2012. "Developments in abrasive flow machining: A review on experimental investigations using abrasive flow machining variants and media." *Proceedings of the Institution of Mechanical Engineers,*

Part B: Journal of Engineering Manufacture. 226 (12): 1951–1962. doi: 10.1177/0954405412462000

[3] Hull, J. B., D. O'Sullivan, A. J. Fletcher, S. A. Trengove, and J. Mackie. 1992. "Rheology of carrier media used in abrasive flow machining." *Key Engineering Materials.* 72–74: 617–626. doi: 0.4028/www.scientific.net/KEM.72-74.617.

[4] Jain, V. K., and S. G. Adsul. 2000. "Experimental investigations into abrasive flow machining (AFM)." *International Journal of Machine Tools and Manufacture.* 40(7): 1003–1021. doi: 10.1016/S0890-6955(99)00114-5

[5] Jain, V. K. 2014. *Advanced Machining Processes*. New Delhi, India: Allied Publishers Private Limited, 58–72. ISBN: 87-7764-294-4

[6] Loveless, T. R., R. E. Williams, and K. P. Rajurkar. 1994. "A study of the effects of abrasive-flow finishing on various machined surfaces." *Journal of Materials Processing Technology.* 47 (1): 133–151. doi: 10.1016/0924-0136(94)90091-4

[7] Perry, W. B. 1985. *Abrasive Flow Machining: Principles and Practices*. Cincinnati, Ohio, USA: Proceedings of Non-Traditional Machining Conference, 121–128.

[8] Petare, A. C., and N. K. Jain. 2018. "A critical review of past research and advances in abrasive flow finishing process." *The International Journal of Advanced Manufacturing Technology.* 97(1):741–782. doi: 10.1007/s00170-018-1928-7

[9] Przyklenk, K. 1986. *Abrasive Flow Machining: A Process for Surface Finishing and Deburring of Workpieces with a Complicated Shape by Means of Abrasive Laden Media*. ASME-PED: Advances in Non-traditional Machining, Vol. 22, 101–110.

[10] Rajeshwar, G., J. Kozak, and K. P. Rajurkar. 1994. *Modeling and Computer Simulation of Media Flow in Abrasive Flow Machining Process* Chicago, USA: Proceedings of the 1994 International Mechanical Engineering Congress and Exposition, 965–971.

[11] Rhoades, L. 1991. "Abrasive flow machining: A case study." *Journal of Materials Processing Technology.* 28(1): 107–116. doi: 10.1016/0924-0136(91)90210-6

[12] Sankar, M. R., V. K. Jain, and J. Ramkumar. 2009. “Experimental investigations into rotating workpiece abrasive flow finishing.” *Wear.* 267(1): 43–51. doi: 10.1016/j.wear.2008.11.007

[13] Williams, R. E, K. P. Rajurkar, and L. Rhoades. 1989. *Performance Characteristics of Abrasive Flow Machining*. Non-Traditional Machining, Dearborn, MI, USA: SME Technical paper, FC 89-806, 898– 906

[14] Williams, R. E., and K. P. Rajurkar. 1992. "Stochastic modeling and analysis of abrasive flow machining." *Journal of Engineering for Industry*. 114(1): 74–81. doi: 10.1115/1.2899761

CHAPTER TWO

INTRODUCTION TO GEARS

The gear is a mechanical component used to transmit motion or power between the two shafts through successive engagements of teeth cut on its periphery. It constitutes an economical means to transmit motion and power with high accuracy. Different types of gears are used in different industries, scientific organizations, and commercial and domestic applications, including automobiles (i.e. cars, trucks, tractors, motorcycles, scooters, etc.), the aerospace industry (i.e. high-speed aircraft engines), the marine industry (i.e. high power high-speed marine engines, navy fighting ships), control systems (i.e. gun, helicopter, tanks, radar applications), earthmoving equipment, home appliances (i.e. washing machine, food mixtures, fans, etc.), toys, gadgets, micro and nanodevices, and the oil and gas industry (i.e. oil platforms, pumping stations, drilling sites, refineries, and power stations). Gears can be classified according to the following criteria (Davis, 2005; Townsend, 2011; Radzevich, 2016):

- **Relative position of the axes of the shafts on which gears are mounted:** Gears mounted on the parallel shafts are manufactured from the cylindrical blanks, therefore they are known as *cylindrical gears*. Spur, single helical, double helical, and herringbone gears belong to this category. Gears mounted on the intersecting shafts are manufactured from the frustum of conical blanks so they are referred to as *conical gears*. Straight bevel, spiral bevel, zero bevel, miter gear, and face gear or crown wheel are examples of conical gears. Gears mounted on non-parallel and non-intersecting axes are called *skew shaft gears*. Rack and pinion, crossed axes helical gears, worm and worm wheel, cylindrical worm gears, single enveloping worm gears, double enveloping worm gears, hypoid gears, spiroid, and helicon gears belong to this category.

- **Shape of the gear:** Circular or non-circular gears. triangular, square, rectangular, pentagonal, elliptical, scroll, lobbed, segmented, and multiple sector gears are some examples of the non-circular gears. They are mostly used where variable velocity ratio, non-uniform motion, interrupted motion with dwell time, and special motion are desired.
- **Size of major diameter:** Gears can be classified as micro-sized (0.1 to 1 mm), meso-sized (1 to 10 mm), and macro-sized (above 10 mm).
- **Location of gear teeth:** Can be classified as external or internal type gears.
- **Profile of gear teeth:** Can be classified as involute, cycloidal, or mixed profile gears.
- **Shape of gear teeth:** Can be classified as straight, inclined, or curved teeth gears.
- **Peripheral velocity:** Can be classified as low velocity (less than 3 m/s) gears, medium velocity (3 to 15 m/s) gears, or high velocity (above 15 m/s) gears.
- **Form of gear profile:** Can be classified as standard gears or profile shifted gears.
- **Pressure angle:** Can be classified as constant pressure angle gears or variable pressure angle gears.
- **Function:** Can be classified as motion transfer gears, power transfer gears, and precision gears for scientific applications.

Figure 2.1 depicts photographs of different types of gears.

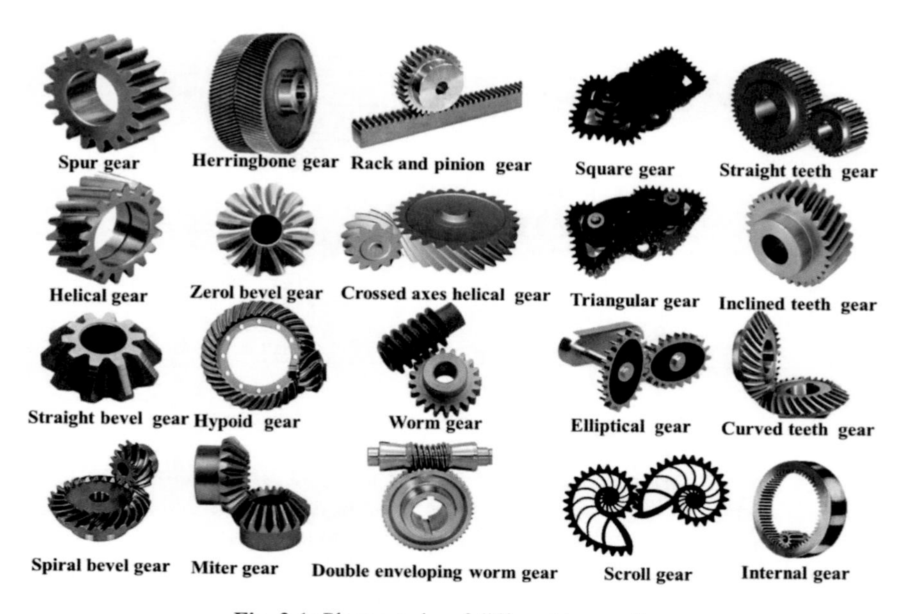

Fig. 2.1: Photographs of different types of gears

2.1 Materials for Gears

The selection of an appropriate material for a gear is an important factor to achieve its desired performance, strength, resistance to wear, impact and fatigue, and service life. It is based on the required shape, size, application, transmission, tolerances, surface finish, and economic aspects. Both metallic and non-metallic materials can be used to manufacture the gears mentioned in table 2.1. Metallic materials can be further classified as ferrous and non-ferrous. Ferrous gear materials include plain carbon steel, cast steel, tool steel, stainless steel, and cast iron whereas copper and its alloys (brass and bronze) and the alloys based on titanium, nickel, aluminum, and magnesium are the commonly used non-ferrous gear materials. Nylon, Teflon, Metlon, Phenolic, Derlin, and olypropylene are the commonly used non-metallic gear materials.

Table 2.1: Different types of gear materials

Metallic		Non-metallic (Plastics)
Ferrous	**Non-ferrous**	
Cast iron	Aluminum alloys	Nylon
Cast steel	Copper alloys (Brass and bronze)	Teflon
Plain carbon steel	Titanium alloys	Derlin
Stainless steel	Nickel alloys	Phenolic
Tool steel	Magnesium alloys	Metlon
	Sintered powder	Polypropylene

2.2 Manufacturing Processes for Gears

The manufacturing processes for gears can be classified as traditional and non-traditional or advanced processes. Traditional processes are further classified according to cylindrical or conical gear manufacturing processes. The manufacturing processes for cylindrical gears are further divided into the following three categories: (i) subtractive type, i.e. gear machining processes, (ii) deformative type, i.e. gear forming processes, and (iii) accretion type. Cylindrical gear machining processes are further classified as the generative processes, which use a generic tool for cutting the gear, and the form cutting processes, which use a tool whose shape corresponds to the shape of the gap between two consecutive teeth to be cut for a particular type of gear. Conical gear manufacturing processes are classified into two categories: (i) generative, and (ii) non-generative. Non-traditional or advanced manufacturing processes for gears are further categorized into four categories, namely (i) subtractive type, i.e. spark erosion machining (SEM), μ-SEM, wire SEM (WSEM), μ-WSEM, water jet machining (WJM), abrasive water jet machining (AWJM), and laser beam machining (LBM); (ii) additive type; (iii) deformative type; and (iv) hybrid type. Table 2.2 illustrates the various manufacturing processes used for gears of each type.

Table 2.2: Different types of gear manufacturing processes (Gupta et al., 2017)

Traditional manufacturing processes for gears			Non-traditional or advanced manufacturing processes for gears
For cylindrical gears		For conical gears	
Subtractive type • Gear hobbing • Gear shaping • Gear planing **Form cutting** • Gear milling • Gear broaching • Gear cutting on a shaper • Shear cutting	**Accretion type** • Casting • Powder metallurgy • Injection molding of plastic gears **Deformative type** • Gear rolling • Gear forging • Gear extrusion • Cold drawing • Stamping and fine-edge blanking	**Generative type** • Face milling • Face hobbing • Using interlocking cutters • Using Revacycle cutters • Shaping using two cutters • Planing generator **Non-generative type** • Template machining • Format machining • Helix form machining • Cyclex machining	**Subtractive type** • Spark erosion machining (SEM) • μ-SEM • Wire SEM (WSEM) and μ-WSEM • Water jet machining (WJM) • Abrasive WJM (AWJM) • Laser beam machining (LBM) **Additive type** • Metal injection molding • Injection compression molding • Micro-powder injection molding • Additive manufacturing **Deformative type** • Hot embossing • Fine blanking **Hybrid type** • Lithographie, Galvanoformung Abformung (LIGA)

2.3 Gear Quality Evaluation Parameters

More than 10 billion gears are manufactured annually and used in various applications in almost all industries. Such a huge demand makes the gear manufacturers focus on producing lightweight and high-quality gears economically that can transmit the required power or motion and simultaneously meet the requirements depicted in fig. 2.2. These requirements can be grouped into the following three categories: (i) **desired gear design characteristics:** minimum size, lightweight, better power and motion transfer characteristics, less maintenance; (ii) **better gear performance requirements:** higher load-carrying capacity, generating less noise and fewer vibrations, higher transmission efficiency; and (iii) **gear service life-enhancing requirements:** better surface quality, and a higher resistance to wear and fatigue (Karpuschewski, 2008). The fitness of a gear to meet these requirements can be evaluated by grouping the evaluation parameters into two groups, namely surface quality and surface integrity.

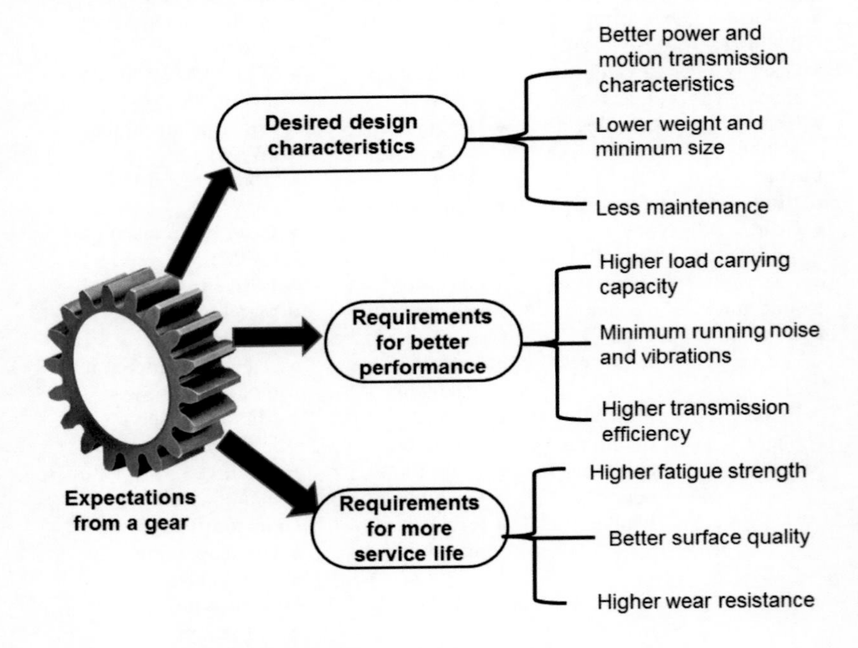

Fig. 2.2: Desired requirements from a gear

2.3.1 Quality Standards for Gears

There are different international standards gear quality. Some standards worth mentioning are the *Deutsches Institut fur Normung* (DIN) of Germany, American Gear Manufacturing Association (AGMA), Japanese International Standard (JIS), Korean Standard (KS) of South Korea, Italian Organization of Standardization (UNI), British Standards Society (BSS), and International Organization for Standardization (ISO) headquartered in Geneva. The most commonly used standards in industry to define gear quality are AGMA 3900, DIN 3962, JIS B1702, and ISO 1328 (Radzevich,

2016; Goch, 2003). Table 2.3 presents the typical applications of gears according to different quality standards.

Table 2.3: Typical applications of gears according to their quality (Jain and Chaubey, 2017)

Designation	**Gear quality standard**				**Typical applications**
	AGMA 2000/2009	**DIN 3962**	**JIS B 1702**	**ISO 1328**	
AA: Ultra-high accuracy	14 or 15	2 or 3	-	2 or 3	Military navigation, precision, instruments, computer
A: High accuracy	12 or 13	4 or 5	0 to 1	4 or 5	Instruments, aircraft engines, turbines, professional cameras, precision machine-tool, speed drives, automotive transmissions, high-speed machinery
B: Medium-high accuracy	10 to 11	6 to 7	2 to 3	6 to 7	
C: Medium accuracy	8 to 9	8 to 9	4 to 5	8 to 9	Appliances, hand tools, pumps, commercial clocks, farm machines, fishing reels, hoists, slow-speed machinery
D: Low accuracy	6 or 7	10 or 1	6 to 7	10 or 11	
E: Very low accuracy	4 or 5	12	8	12	

2.3.2 Gear Surface Quality

The surface quality of a gear is evaluated in terms of errors in its microgeometry (i.e. form and location errors), flank surface topography, functional performance parameters (which are determined by single and double flank roll testing), surface roughness, and wear characteristics. The poor surface quality of a gear is a major source of errors in motion transfer, excessive wear, generation of noise and vibrations, and premature failure (Davis, 2005). Microgeometry errors and the surface roughness of a gear are the most dominant factors causing gear noise generation. Both these factors contribute approximately 30% to the overall production cost of a gear (Townsend, 2011). Achieving minimum running noise, maximum load carrying, enhanced service life and better tribological fitness, higher wear resistance, and better operating performance from a gear require a reduction in its microgeometry errors and surface roughness using the appropriate gear finishing process(es). Figure 2.3 depicts the different aims of a gear finishing process.

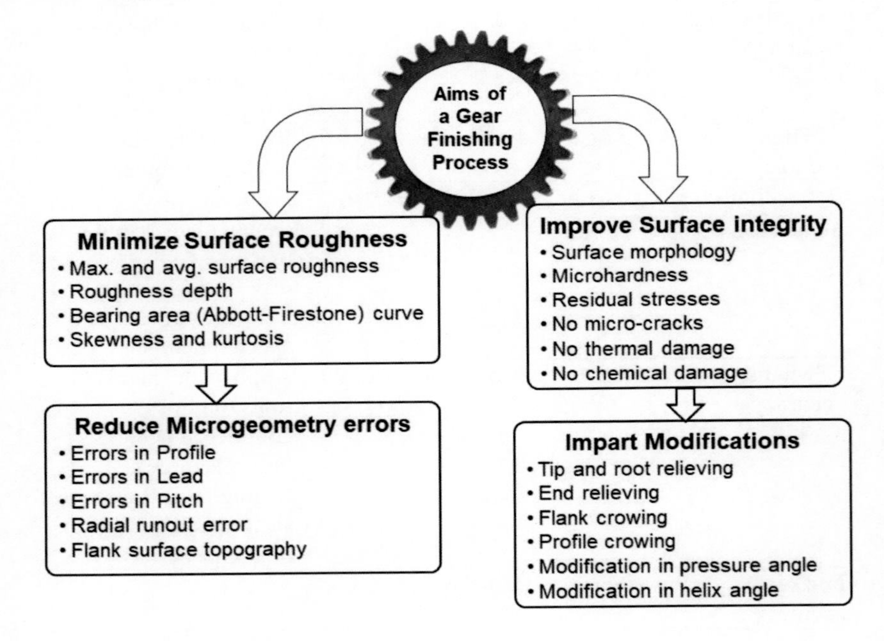

Fig. 2.3: Different aims of a gear finishing process

2.3.2.1 Microgeometry Errors

Microgeometry errors of gear are associated with the form and location errors of its teeth. Figure 2.4 illustrates different components of errors in gear microgeometry and their effects on its various performance aspects. *Form errors* have two components, namely errors in profile and errors in the lead of a gear. *Total profile error* 'F_a' is a combination of deviations and errors in the form and slope of an involute profile of a gear. It is evaluated perpendicular to the involute profile by tracing the probe from root to tip, preferably at the middle of the flank width of the selected teeth of a gear. It significantly affects the gear noise generation characteristics. It is the sum of two components: (i) *Profile form error* 'ffa' (the difference between nominal involute form and actual involute form); and (ii) *Profile angle error* 'fha' (the difference between a nominal involute angle and actual involute angle). The *total lead error* 'F_β' is a combination of errors in the form and slope of flank surface of the selected teeth of a gear. Evaluated by tracing the probe at the pitch line along the face width of a gear, it has a major influence on the load-carrying capacity of a gear. It is a sum of two components: (i) *Lead form errors* '$ff\beta$' (the difference between nominal lead form and actual lead form); and (ii) *Lead angle errors* '$fh\beta$' (the difference between nominal helix angle and actual helix angle). Errors associated with the pitch and runout of all the teeth of a gear are referred to as *location errors*. They affect the motion transfer characteristics and transmission efficiency of a gear and are evaluated by touching the probe on the middle points of both flanks of all the teeth of a gear at their pitch points along its pitch circle. *Pitch error* describes the middle location of all right and left flanks concerning each other. It has three components, namely: (i) *Single pitch error* 'f_p' is the maximum absolute value among all the individual single pitch deviations (i.e. $f_p = max\ |f_{pi}|$) where individual single pitch deviation 'f_{pi}' is the algebraic difference between the actual pitch and the corresponding theoretical pitch, (ii) *Adjacent or successive or pitch-to-pitch error* 'f_u' is the maximum difference between two consecutive pitches measured on the specified flank (i.e. either the right or the left) of all the teeth of a gear, and (iii) *Total cumulative pitch error or total index error* 'F_p' is the difference between the summation of the theoretical values of all the pitches and the summation of the actual values of all the pitches taken over all the teeth of a gear. *Radial runout* (determined from microgeometry measurement) 'F_r' describes the radial location of all teeth with respect to the pitch circle. It is the maximum difference between the actual radial positions of all teeth measured with respect to their nominal radial position.

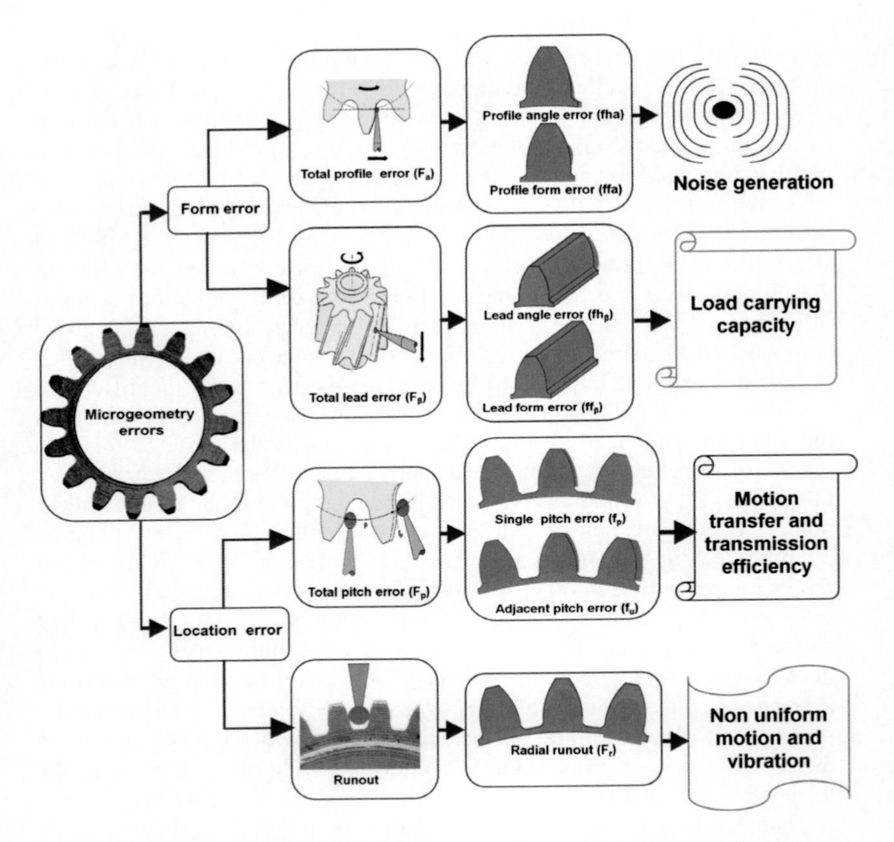

Fig. 2.4: Components of microgeometry errors of a gear and their effects

2.3.2.2 Flank Surface Topography

The topography of the gear tooth flank surface is a three-dimensional graphical representation depicting the combined effects of errors in its profile and lead and shows the differences between actual and theoretical flank surfaces through peaks and valleys. It indicates the presence of nicks, burr, peaks, and valleys on the actual flank surface of a gear tooth, which can help in understanding the wear and service life of a gear. Flank topography data for a gear are collected by dividing its actual flank surface along with its profile and the flank width in a certain number of discrete points thus forming a grid. Measurements are taken on each point of this grid and are joined by straight lines along the lead and profile directions. Actual and theoretical flank surfaces are represented in different colors whereas the amount of deviation between the actual and theoretical flank surfaces is represented by vertical lines at each grid point whose color is different than that of the actual and theoretical flank surfaces (Goch, 2003).

2.3.2.3 Functional Performance Parameters

The functional testing of a gear simulates its actual working conditions by meshing it with a master gear and recording variations in the center-to-center distance as they rotate. It can be either single flank roll testing or double or dual flank roll testing depending on the type of contact between the flank surfaces of the test and master gears. Figures 2.5a and 2.5b depict the working principles of single flank roll testing and double flank roll testing, respectively. Double flank roll testing is used to determine total composite error 'F_i', tooth-to-tooth composite error 'f_i', and radial runout 'F_{rf}', whereas single flank roll testing is used to determine the parameters most closely related to typical gear noise. This includes total transmission error, tooth-to-tooth transmission error, effective profile error, adjacent pitch error 'f_u', and total or cumulative pitch error or total index error 'F_p'. The ability of single flank roll testing to determine total cumulative pitch error is an important aspect because a gear with radial runout error will certainly have total cumulative pitch error, but a gear with total cumulative pitch error may not necessarily have radial runout error. Radial runout occurs in a gear with a bore or locating surface that is eccentric from the pitch circle of the gear teeth. It can be a large total composite error if inspected during double flank roll testing.

Total transmission error is the difference between the actual position of an output (or driven) gear and the position it would occupy if the mating gears were perfectly conjugate. It is a measure of the *profile conjugacy* of the mating gears expressed in radians. *Tooth-to-tooth transmission error* is the variation in total transmission error of the *gear meshing frequency* (product of the number of gear teeth and its rotational speed in revolutions per second), and is mainly the consequence of profile errors and single pitch error. *Effective profile errors* indicate some non-uniform motion that the gear is likely to transmit, which causes a shaking of the gear-supporting structure and generates noise. A perfect involute profile is desirable for noise control in light-load applications. Gear teeth profiles are often modified to obtain a compromise between the load carrying capacity and the smoothness of the transmitted motion. Such modifications produce predictable intentional variations in single flank roll testing; for example, providing root relief and tip relief leads to a parabola-like motion curve while errors in the gear pressure angle cause a saw-tooth motion curve. *Tooth-to-tooth composite error* 'f_i' is the maximum variation in the center-

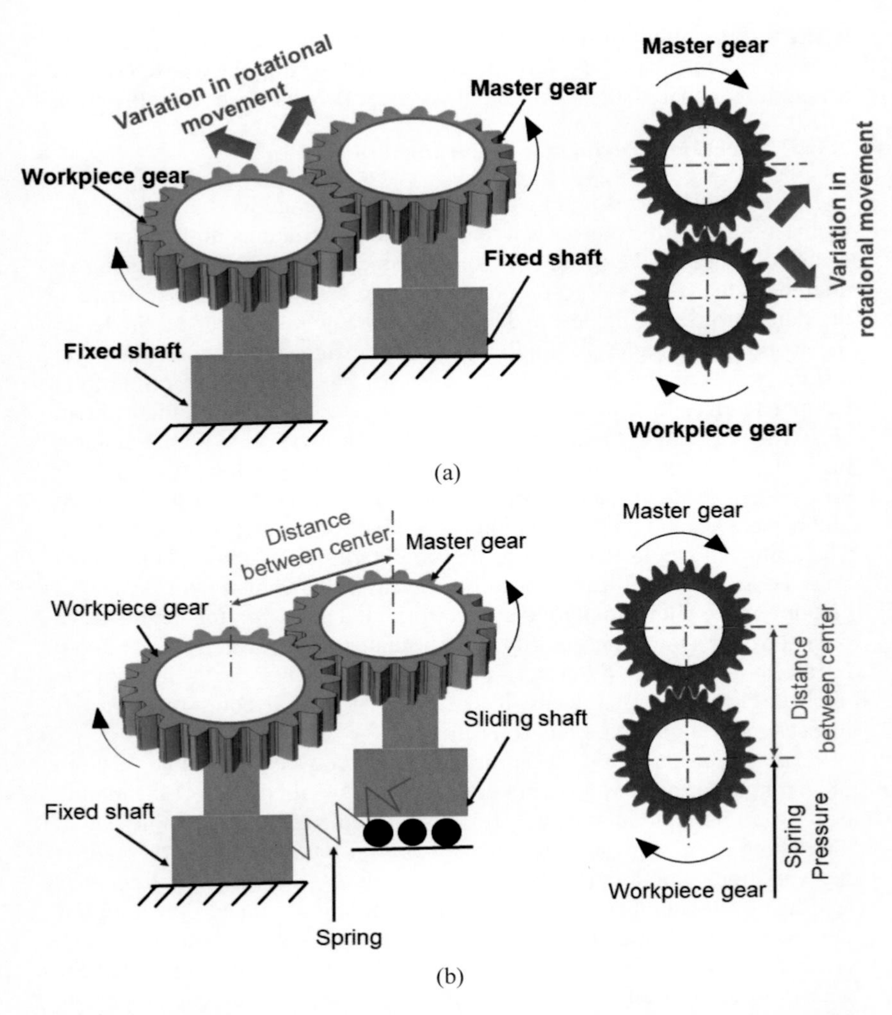

Fig. 2.5: Working principles of (a) single flank roll testing, and (b) double or dual flank roll testing

to-center distance between the test and master gears per revolution per tooth of the test gear (i.e. 360/number of teeth in the test gear). It includes the effects of errors in the profile, pitch, tooth thickness, and tooth alignment in both the test and master gears. *Total composite error* 'F_i' is the total change in the center-to-center distance in one complete revolution of the test gear.

It is a combination of the radial runout with tooth-to-tooth composite error. *Radial runout* (determined from functional testing) 'F_{rf}' of a test gear is the difference between the maximum and the minimum radial distance from the gear axis as observed after removing the short-term or undulation pitch deviations and analysing the long-term sinusoidal waveform (Arteta et al., 2013).

2.3.2.4 Surface Roughness Parameters

Surface roughness parameters are very important for components subjected to fluctuating stresses such as gears, shafts, pistons, camshafts, bearings, etc. Surface roughness affects the service life and operating performance of a gear. A higher surface roughness causes less available contact area with the meshing gear and frequent breakages of the roughness peaks, which result in the faster wear of a gear by pitting, micro-pitting, scuffing, abrasive wear, and adhesive wear. Average surface roughness 'R_a' is the arithmetic average of the absolute values of roughness profile deviations from the mean line over one sampling length 'l_r'. It is the most commonly used parameter for the analysis of surface roughness. Figure 2.6a shows its concept and it is mathematically expressed by Eq. 1:

$$R_a = \frac{1}{l_r}\int_0^{l_r} |Z(x)|\, dx \qquad (1)$$

Single roughness depth 'R_{zi}' is the vertical distance between the highest peak and the deepest valley within a sampling length. Mean roughness depth 'R_z' is the arithmetic average of the single roughness depths of consecutive 'n' number of sampling lengths. It is conceptually depicted in Fig. 2.6b and mathematically expressed by Eq. 2. Maximum surface roughness 'R_{max}' the largest single roughness depth within the evaluation length, as shown in Fig. 2.6b.

$$R_z = \frac{1}{n}(R_{z1} + R_{z2} + \cdots \ldots \ldots . + R_{zn}) \qquad (2)$$

.

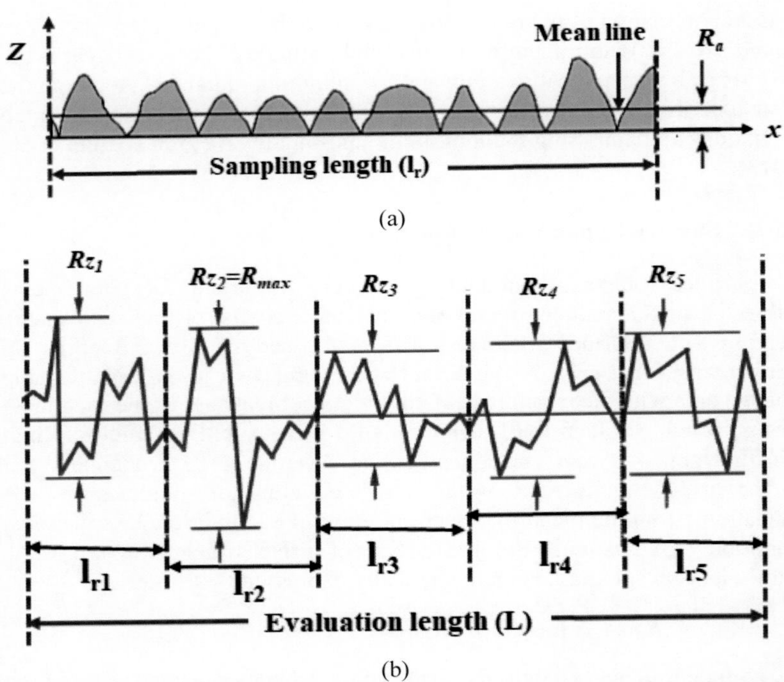

Fig. 2.6 Graphical representation of the surface roughness parameters: (a) average surface roughness 'R_a'; (b) mean roughness depth 'R_z', and maximum surface roughness 'R_{max}'

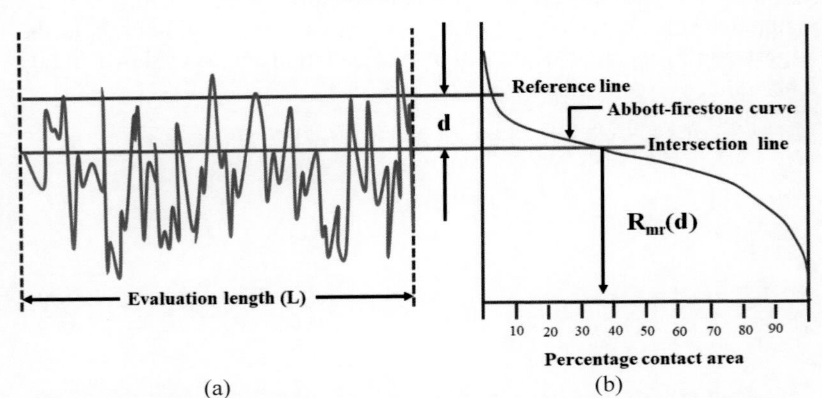

Fig. 2.7: Concepts of (a) bearing area ratio 'R_{mr}'; and (b) Abbott-Firestone or bearing area curve

The bearing area ratio 'R_{mr}' is the ratio expressed in percent of the material-filled length to the evaluation length at the profile section level *'d'*, which is the distance between the evaluated intersection line (shown as a red line in Fig. 2.7a) and the specified reference line (shown in in Fig. 2.7a). The Abbott-Firestone or bearing area curve shows the material ratio 'R_{mr}' as a function of the profile section level *'d'*, as shown in Fig 2.7b. *Skewness* 'R_{sk}' indicates the asymmetry of distribution of the surface asperities within the sampling length. It is an important parameter because it gives information about the morphology of the surface texture. Its positive value indicates a distribution of higher peaks on gear flank surfaces, and its negative value means that the flank surfaces of a gear have good bearing properties and pores that are very good for lubrication purposes. *Kurtosis* 'R_{su}' is a measure of the sharpness of the peaks of distribution of the surface asperities within the sampling length, and values in excess of three indicate the flank surfaces of the gear have good bearing properties.

2.3.2.5 Wear Characteristics

The wear of a gear involves the removal or transfer of material from its flank surfaces due to its rolling and sliding contact with the flank surfaces of its meshing gear. It starts during the initial running of the mating gears and increases progressively with their continuous use. It is significantly affected by the lubrication oil used and its contamination. There are four mechanisms of wear of a gear, namely adhesive, abrasive, polishing, and scuffing. The *adhesive wear* of a gear takes place due to the formation of micro-welds between the flank surfaces and the mating gears and their subsequent rupture. *Abrasive wear* occurs due to debris, microchips, dust, and other contaminants entering its lubrication system. *Polishing wear* happens due to the gear's chemical reaction with its lubricant. It is visible in the scratch marks in the sliding direction of mirror-finished flank surfaces. *Scuffing* is a severe form of gear wear that occurs due to plastic deformation and deep scratches on its flank surfaces increasing the coefficient of friction and temperature of the meshing gears (Saeidi et al., 2017).

2.3.3 Gear Surface Integrity

Surface integrity is the description and control of many possible metallurgical and other alterations (i.e. mechanical, thermal, chemical, and electrical damage) produced within 0.5 mm under the visible manufactured surface. This includes metallurgical and other alterations on the material properties, quality, and performance of the manufactured surface during its service life. Surface integrity significantly influences the performance of those components that are subjected to dynamic loading, fluctuating loading and fatigue, friction, wear, and severe corrosive environments such as acidic, alkaline, saline, or reactive gaseous (Astakhov, 2010). For example, operating stresses are controlled by the fatigue characteristics of a component in case of dynamic loading, and fatigue failures often initiate at or just below the surface of a component. Similarly, in the case of stress corrosion resistance, the surface condition of a component is a primary factor in determining its susceptibility to attack and its possible subsequent failure. The manufacturing, finishing, and storing processes of a product affect its surface integrity. Although all the manufacturing processes significantly affect the workpiece material and its properties, most of the currently available data are about the material removal (i.e. machining) processes because the final process for most engineered components is a finishing process of subtractive nature.

The study of gear surface integrity involves an examination and analysis of surface morphology to get an idea of the mechanism of material removal, mechanical damage (i.e. hardness alterations, plastic deformation, residual stresses, and microcracks), thermal damage (heat affected zone, microstructure changes, recrystallization, recast material, and redeposited material), and chemical damage (i.e. intergranular attack, alloy depletion due to selective etching, corrosion, stress corrosion, contamination, and embrittlement). Optical microscopy (OM), scanning electron microscopy (SEM), and atomic force microscopy (AFM) with proper magnification can be used to study surface morphology (at lower magnification) and microstructure (more than 1000 magnification). Residual stresses are determined by indirect methods only, such as X-ray diffraction (XRD), neutron diffraction (ND), synchrotron diffraction (SD), electron diffraction technique, ultrasonic technique, magnetic technique, and Barkhausen noise analysis (BNA). Hardness alterations can be determined by evaluating change in microhardness using Vickers or Knoop microhardness measuring equipment. All these examination techniques require the preparation of samples following standard procedures.

Since a gear is subjected to fluctuating and dynamic loading, the surface integrity of its flank surfaces significantly affects its operating performance, service life, wear, and fatigue resistance. Surface morphology and microhardness affect the load-carrying capability and wear resistance of gear while residual stresses affect its fatigue strength; therefore, they are most commonly used to evaluate the surface integrity of the flank surfaces of a gear and are described in the following sections.

2.3.3.1 Surface Morphology

The morphology of a manufactured surface can be viewed using a magnification of up to 1000x by OM, SEM, or AFM. It can be used to determine the mechanism of material removal or wear, and to view the presence of the marks by a cutting tool, flow path of abrasives, scratches, burrs, microchips, microcracks, debris, pits, voids, tears, recast, redeposited material, and heat-affected zones (HAZ).

2.3.3.2 Microhardness

Microhardness is determined by applying a very small load (maximum value up to 1000 g) on the surface of a manufactured component, thus producing a negligible indentation on it. This measures the hardness from the surface to the core on a heat-treated, manufactured, and finished component. Microhardness studies of a gear help determine the changes in its flank surface hardness before and after manufacturing, finishing, and heat treatment processes. It helps identify the effects of the recast layer and HAZ on the hardness of flank surfaces of a gear and determine the thickness of the recast layer.

2.3.3.3 Residual Stresses

Residual stresses present the internal stress distribution inside a material in the absence of an external loading. They may be compressive or tensile in nature. Almost all manufacturing processes introduce residual stresses to a manufactured component. Fatigue failure is a common phenomenon in gears because they are subjected to variable loads. The fatigue strength of flank surfaces can be improved by inducing compressive residual stress through shot peening, laser shock peening, or laser punching processes.

The traditional manufacturing processes for gears do not yield acceptable gear quality thus the gears fail to meet the requirements of the end-users; for example, gears manufactured by hobbing and shaping processes have tool marks and scallops on their flank surfaces. Post-manufacturing heat treatment to improve gear hardness also adversely affects the quality of the gears. Therefore, an appropriate gear finishing process is necessary to ensure a better surface quality of a gear by eliminating the imperfections/discontinuities (i.e. small pits, burrs, scratches, cut marks, etc.) that occurred during manufacturing and heat treatment.

2.4 Traditional Finishing Processes for Gears

The selection of an appropriate finishing process for gears depends on factors such as the hardness of the gear material, the required surface finish and surface integrity, and the process productivity. Gear grinding, gear lapping, gear honing, gear shaving, gear burnishing, and gear skiving are traditional finishing processes. Table 2.4 presents the capabilities, applications, and limitations of these processes in a comparative form, and the following paragraphs describe their working principles.

Table 2.4: Comparison of different traditional finishing processes for gears (Jain and Petare, 2017)

Process	Capabilities	Applications	Limitations
Gear Grinding	● It can finish case-hardened gears to a hardness value of up to 92 HRC. ● Reduces errors in the profile, lead, and pitch of all types of gears. ● It can make profile modifications.	● It can finish all types of cylindrical and conical gears with a large values of face width, module, and pressure angle. ● For finishing those gears that require better accuracy in profile and lead. ● For mass finishing of different sizes of a particular gear.	● Grinding burn marks on the flank surface damages the surface integrity of the ground gears. ● Generation of grind lines on the flank surface of a gear leads to generation of noise and vibrations. ● Requires redressing of the grinding wheel and a continuous coolant supply. ● Requires a formed grinding wheel. ● Higher operating and maintenance costs.
Gear Lapping	● It can correct only minor irregularities in the involute profile, lead angle, total pitch, and runout. ● It can correct any minor distortion caused by the heat treatment of a gear.	● For finishing all types of cylindrical and conical gears. ● For finishing the hobbed, shaved, and hardened gears. ● Used when higher dimensional accuracy and excellent surface finish are required	● Increasing lapping time adversely affects the form errors of a gear. ● Applicable for finishing mating gears only and when the lapped pair cannot be interchanged.

Process	Capabilities	Applications	Limitations
Gear Honing	● Generation of a crosshatch lay pattern on the flank surface of a finished gear, which improves lubrication oil retention capability. ● Can correct the distortions caused by heat treatment. ● Can do crowning.	● For finishing the hardened external and internal spur and helical gears requiring better surface finish. ● For gears having a higher contact ratio, smaller pressure angle, and longer addendum. ● It can improve minor errors in gear microgeometry.	● Limited life of the honing tool. ● Slow process. ● Increasing finishing time increases tool shape error.
Gear Shaving	● It can reduce microgeometry errors and eccentricity. ● Effectively corrects the spacing error. ● It can reduce around 80% of the manufacturing errors of a gear.	● For finishing external cylindrical and worm gears. ● For mass finishing and form error correction of the case-hardened gears. ● It can improve minor thermal distortions in a gear. ● For mass finishing of different sizes of a particular gear.	● Cannot finish the gears harder than 40 HRC. ● Cannot be used for finishing the internal gears. ● Higher tool wear. ● Shaving cutter is very costly.
Gear Burnishing	● It is a cold working process that improves the surface integrity and fatigue life of the gears. ● Less finishing time compared to other traditional finishing processes for the gears.	● Only for finishing the helical gears.	● Produces localized stress marks and non-uniformity on the flank surface of a gear. ● Can be used only for those gears made of soft or unhardened materials.

Process	Capabilities	Applications	Limitations
Gear Skiving	● Highly accurate and faster finishing process for the gears harder than 92 HRC. ● Lubricating oil retaining capacity of skived gears is better than that of ground gears. ● Machining and finishing are performed in one setting, hence much less chance of radial runout.	● For finishing of the external and internal spur and helical gears. ● Finishing of the internal pinion on armature shafts and shoulder gears is only possible with skiving.	● Cutting tool has fixed geometry and is unable to modify tooth profile. ● Cutter is very costly. ● Chip removal is a problem.

2.4.1 Gear Grinding

Heat treatment processes tend to thermally distort the gears thus deteriorating the accuracy of their teeth. This necessitates hard finishing of the gear teeth. Gear grinding is an effective finishing process for all types of cylindrical and conical gears made of hard or hardened (i.e. heat-treated) materials having a hardness value of up to 92 HRC. It can correct the thermal distortions and improve the surface finish and microgeometry of the gears. It uses a properly formed and dressed grinding wheel that finishes the gear teeth by abrading the fine abrasive particles. It can be classified as form or non-generative grinding or generative grinding according to the shape of the grinding wheel used. Figure 2.8 depicts a schematic of different forms of the gear grinding process. In the generative grinding process, the workpiece gear is run in mesh with two dish-shaped grinding wheels for the spur gear (fig. 2.8a) or a cup-shaped grinding wheel for the spiral bevel gear (fig. 2.8b). The relative motion between the workpiece gear and the grinding wheel finishes the flank surfaces of the gear teeth. It provides accurate tooth spacing, concentricity, and profile. Non-generative or form grinding (figs. 2.8c) uses a circular grinding wheel whose width is 0.25 to 0.50 mm less than the gap between two consecutive teeth of the workpiece gear. When the grinding wheel rotates about its axis, the flanks of two consecutive teeth are ground along with the line contact between it and the teeth being ground.

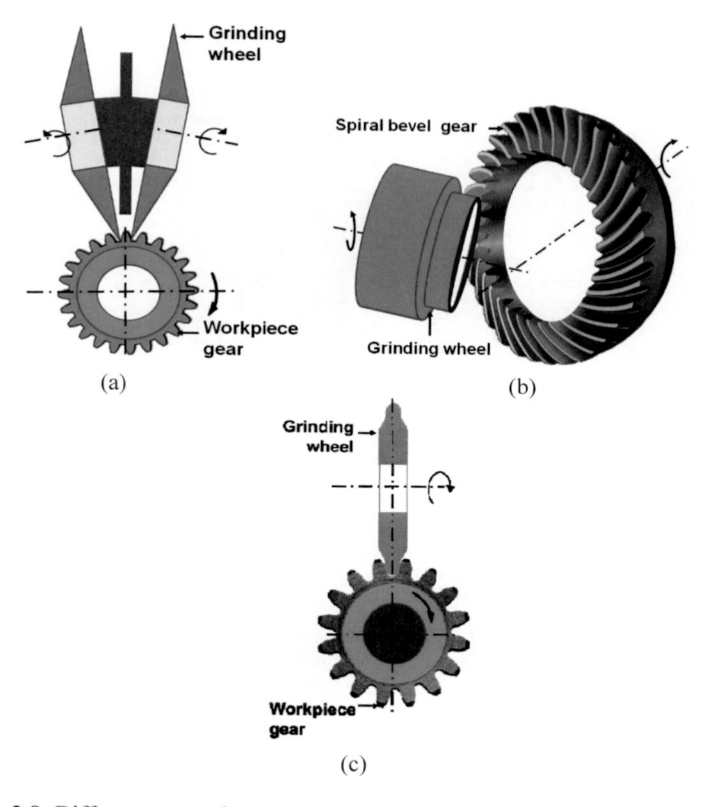

(a) (b)

(c)

Fig. 2.8: Different types of gear grinding: (a) generative grinding of spur gear by the dish-shaped grinding wheel; (b) generative grinding of spiral bevel gear by the cup-shaped grinding wheel; and (c) non-generating grinding of a spur gear using formed grinding wheel

2.4.2 Gear Lapping

Gear lapping simultaneously finishes the mating gears made of hard or hardened materials by running them together in the presence of a lapping medium that is continuously supplied under pressure as depicted in fig. 2.9. The lapping medium is a chalky paste consisting of the abrasive particles between 300 and 900 mesh (i.e. abrasive particle diameter ranging from 17 to 51 μm) and a carrier fluid. Aluminum oxide, silicon carbide, boron carbide, and diamond powder are the most commonly used abrasives. A reduction in the noise level between the mating gears is a criterion for successful lapping.

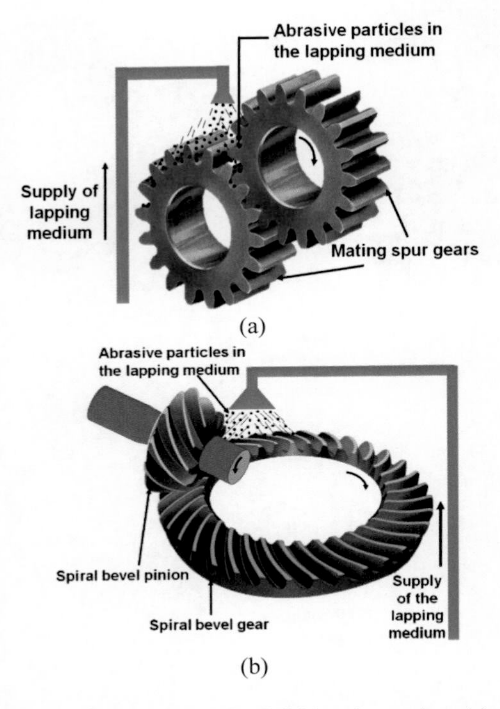

Fig. 2.9: Schematic of gear lapping for finishing the mating: (a) spur gears; (b) spiral bevel gears

2.4.3 Gear Honing

Gear honing uses an abrasive-impregnated helical gear-shaped tool which is run in tight mesh with the workpiece gear in a cross-axis relationship, as shown in fig. 2.10. The workpiece gear is driven by the honing tool at high speeds (up to 1,000 sfpm) and simultaneously traversed back and forth across the honing tool parallel to the axis of the workpiece gear to ensure the finishing of its entire face width. Conventional honing oil is used as a coolant, lubricant, and medium to flush away the wear debris. Gear honing can be used for finishing the hardened external and internal spurs and helical gears. In external gear honing (fig. 2.10a), the workpiece gear and honing gear mesh with each other in a cross-axis arrangement with a constant center distance. Gear teeth are finished from the root to the tip in the direction of the rotation of the honing gear. In the internal gear honing process (fig. 2.10b), a small external helical gear that has cross-axes meshing with the workpiece internal gear held between the centers of a honing machine is used as a honing tool. The workpiece gear is driven in both clockwise and anti-clockwise directions during the honing cycle to ensure the finishing of the left and right-hand flanks of its teeth.

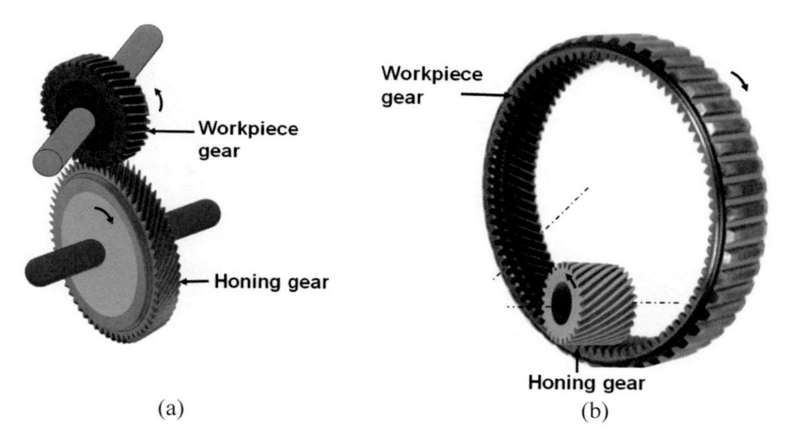

Fig. 2.10: Concept of gear honing process to finish: (a) external gear; and (b) internal gear

2.4.4 Gear Shaving

The gear shaving process removes small amounts of material from the workpiece gear to correct errors in its profile, pitch, helix angle, and eccentricity and improves its surface finish by using a specially shaped shaving cutter similar to a helical gear with serrated cutting edges. Figure 2.11 depicts the schematic of the gear shaving process. The shaving cutter and workpiece gear are pressed to make proper meshing contact, and the center distance between them is reduced incrementally. The shaving cutter rotates in a tight mesh, with the workpiece gear either in parallel or cross-axis arrangement. This meshing is flexible to obtain any desired value of the cross-axes angle in a range from 10 to 15°, which usually yields a better accuracy and surface finish of the gear teeth. Rotary motion between the workpiece gear and shaving cutter is performed in both directions (i.e. clockwise and anticlockwise) during the finishing cycle and material removal takes place in the form of a hair-like chip (Radzevich, 2016). The performance of the gear shaving process depends on factors such as workpiece gear material, its geometry, shaving allowance, the shaving cutter and its material, and parameters related to the shaving process and its machine.

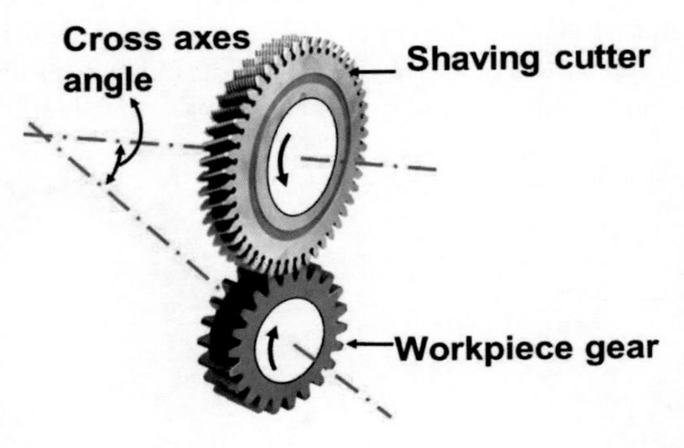

Fig. 2.11: Schematic of the gear shaving process

2.4.5 Gear Burnishing

Gear burnishing is a cold rolling process in which the workpiece gear is rolled under pressure while meshing with an accurately ground, polished, and hardened master gear (also known as burnishing gear) of high accuracy and surface finish, as shown schematically in fig. 2.12. As the burnishing gear (driver) starts engaging with the workpiece gear (driven), the sliding action occurs along the line of action from the top of the burnishing gear tooth towards its pitch point on the approach side, and from its pitch point towards the root of the same tooth on the recession side. These changes in sliding action compress the workpiece gear tooth from the root towards the pitch point on the approach side, and from the pitch point towards the tip on the recession side. They also cause more material to be displaced on the recession side compared to the approach side, approximately in the ratio of 3:1. The workpiece gear is finished by smearing off the minute irregularities from its tooth flank surfaces by the cold plastic deformation along the line of contact, which also improves the fatigue strength of the workpiece gear by inducing compressive residual stresses in it.

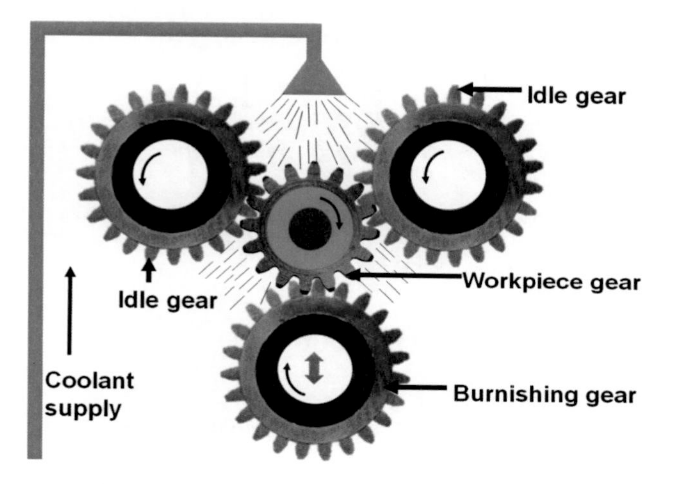

Fig. 2.12: Arrangement of workpiece and burnishing gears in gear burnishing process

2.4.6 Gear Skiving

Gear skiving does both rough machining and finishing of external and internal spur and helical gears by rotating the carbide gear cutter at a very high speed. Continuous axial feed is also provided to the rotating gear cutter parallel to the axis of the workpiece gear blank. Figure 2.13 shows the arrangement of the workpiece gear and hob cutter in the gear skiving process. The front side of the hob cutter has sharp cutting edges that cut the teeth on the workpiece gear blank while meshing and finishing it at the same time. More teeth are engaged per unit time in this process, and it produces highly accurate gear in very little time. This makes gear skiving a highly efficient process, rendering it an effective alternative for finishing the internal gears manufactured by gear broaching and gear shaping processes. It is useful for a medium volume production of hardened gears (Martin et al., 2017).

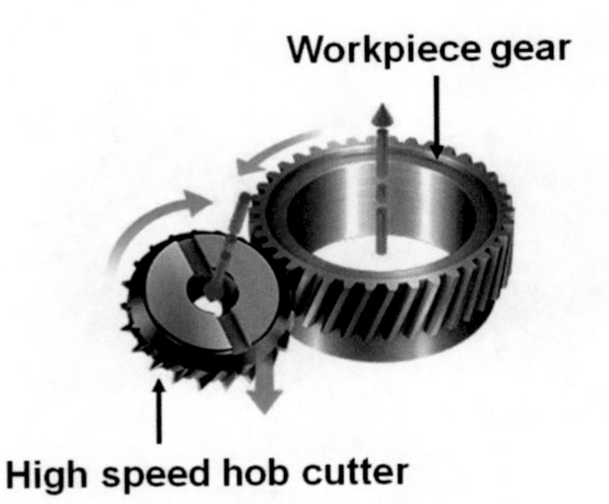

Fig. 2.13: Schematic of the gear skiving process

2.5 Summary of the Limitations of Traditional Finishing Processes for Gears

The following are the major limitations of the traditional finishing processes for the gears:

- Gear grinding results in grinding burn, which damages the surface integrity of the ground gears and leads to premature gear failure. The generation of transverse grind lines on the ground gear teeth flank surfaces leads to the generation of noise and vibrations during use of the ground gear.
- Gear lapping finishes the mating gears in a pair that cannot be interchanged with the members of any other similar pair.
- The gear honing tool has a limited life, and increases in honing time cause increases in tool shape errors.
- Gear shaving can finish either unhardened gears or gears with a hardness value up to 40 HRC only. It leaves a step mark at the end of the involute profile of the gear tooth, which causes excessive noise and vibrations. An increase in shaving time causes the removal of more material from the pitch points of gear tooth flank surfaces, which affects transmission quality.
- Gear burnishing is applicable for finishing unhardened gears but is unable to reduce errors in tooth profile, lead, pitch, and radial runout. It results in localized surface stresses and non-uniform surface characteristics, which cause noise and vibrations.

- Gear skiving is applicable for finishing the internal and external cylindrical gears but is unable to modify the gear tooth profile due to the fixed geometry of the cutter.

It can be concluded from the above-mentioned limitations that none of the traditional finishing processes for gears can provide a techno-commercial solution for finishing all types of gears made of soft, hard, hardened, and unhardened materials. Therefore, there is a strong need to explore and develop a highly productive, easy-to-operate, economic, and sustainable advanced finishing process for gears. This book is an attempt to explore the AFF process for imparting a high-quality finish to spur and bevel gears in an economic and productive manner.

References

[1] Astakhov, V. P. 2010. "Surface Integrity – Definition and Importance in Functional Performance." In *Surface Integrity in Machining*. London, UK: Springer, 1-35. ISBN: 978-1-84882-874-2

[2] Arteta, M. P., J. S. Mazo, R. A. Cacho, and G. A. Arjol. 2013. "Double flank roll testing machines intercomparison for worm and worm gear." *Procedia Engineering*. 63: 454-462. doi: 10.1016/j.proeng.2013.08.231

[3] Davis, J. R. 2005. *Gear Materials, Properties, and Manufacture*, Ohio, USA: ASM International. ISBN: 978-0-87170-815-1

[4] Goch, G. 2003. "Gear Metrology." *CIRP Annals*. 52 (2): 659-695. doi: 10.1016/S0007-8506(07)60209

[5] Gupta, K., N. K. Jain, and R. Laubscher. 2017. *Advanced Gear Manufacturing and Finishing: Classical and Modern Processes*. London, UK: Academic Press. ISBN: 978-012-80-4506-0 (eBook). doi: 10.1016/B978-0-12-804460-5.00008-0

[6] Jain, N. K., and S. K. Chaubey. 2017. "Review of miniature gear manufacturing." In *Comprehensive Materials Finishing*. Oxford, UK: Elsevier, 504-538. doi: 10.1016/B978-0-12-803581-8.09159-1

[7] Jain, N. K., and A. C. Petare. 2017. "Review of gear finishing processes." In *Comprehensive Materials Finishing*. Oxford, UK: Elsevier 93-120. doi: 10.1016/B978-0-12-803581-8.09150-5

[8] Karpuschewski, B., H. J. Knoche, and M. Hipke. 2008. "Gear finishing by abrasive processes." *CIRP Annals*. 57 (2): 621-640. doi:10.1016/j.cirp.2008.09.002

[9] Pathak, S, N. K. Jain, and I. A. Palani. 2019. *Finishing of Conical Gears by Pulsed Electrochemical Honing*. Newcastle upon Tyne, UK: Cambridge Scholars Publishing, ISBN: 1-5275-3366-2

[10] Radzevich, S. P. 2012. *Dudley's Handbook of Practical Gear Design and Manufacture*. New York, USA: CRC Press. doi:10.1201/9781315368122

[11] Saeidi, F., M. Parlinska-Wojtan, P. Hoffmann, and K. Wasmer. 2017. "Effects of laser surface texturing on the wear and failure mechanism of grey cast iron reciprocating against steel under starved lubrication conditions." *Wear*. 386-387: 29-38. doi:10.1016/j.wear.2017.05.015

[12] Townsend, D. P.1992. *Dudley's Gear Handbook*, New Delhi, India: McGraw-Hill Inc. ISBN:10: 0071077367

CHAPTER THREE

PAST WORK REVIEW

This chapter presents reviews of the past work relevant to (i) gear finishing by the abrasive flow finishing (AFF) process, (ii) use of laser texturing different materials for improving the performance of some engineering components, and (iii) study on reduction of gear noise and vibrations by different finishing processes. It also describes the existing gaps based upon this review, the study's objectives, and the investigation methodology used to achieve them.

3.1 Past Work on Gear Finishing by the AFF Process

Only the three references briefly described in the following paragraphs are available on gear finishing by the AFF process. Table 3.1 presents a summary of them.

Xu et al. (2014) used the AFF process to finish helical gears and studied the influence of the number of finishing cycles on their surface roughness. They reported that the (i) surface roughness values at the left flank, right flank, and addendum of the helical gears before the AFF process were 1.42, 1.10, and 2.73 μm, respectively, and were reduced to 0.22, 0.21, and 1.75 μm, respectively, after finishing by the AFF process; and (ii) tooth contact stiffness, fatigue strength, and surface quality of the AFF-finished helical gears improved with very little finishing time. Venkatesh et al. (2014) used the AFF process to finish the bevel gears made of EN-8 steel by varying the extrusion pressure, size of the abrasive particles, finishing time, and flow rate of the AFF medium. They studied the interaction of the abrasive particles on bevel gear flank surfaces using the computational fluid dynamics (CFD) approach, and reported that (i) extrusion pressure, size of the abrasive particles, and finishing time are significant parameters; (ii) extrusion pressure contributed 73% to the surface finish improvement and 83% to the material removal; and (iii) surface morphology improved with an increase in the extrusion pressure. Venkatesh et al. (2015) used ultrasonically assisted AFF (UA-AFF) to finish the straight bevel gears made of EN-8 steel using a mixture of natural rubber and SiC abrasive

Table 3.1: Summary of past work on gear finishing by the AFF process

Researchers (Year)	*Finished component (their material)*	*Process*	*Composition of the AFF medium*	*Used process parameters and their values*	*Responses*
Xu et al. 2013	Helical gears (Details not available)	AFF	Polymer, SiC abrasive particles, and thinner	• Number of strokes: 5,10, 15, 20, 25 • Abrasive particle size: 74 (μm) • Viscosity of the AFF medium: 69 (kPa.s) • Reduction ratio: 0.36 • Extrusion pressure: 1.3 (MPa)	• Surface roughness
Venkatesh et al. 2014	Bevel gear (EN-8 steel)	AFF	Neoprene rubber, SiC abrasive particles	• Finishing time: 5, 7, 9 (min) • Abrasive particles size: 75, 100, 150 (μm) • Extrusion pressure: 2, 3, 4 (MPa) • Flow rate of the AFF medium: 567, 796, 995 (cm^3/min) • Viscosity of the AFF medium: 0.51 (kPa.s) • Concentration of abrasive particles: 50 (wt %) • Temperature of the AFF medium: 32 ± 2 °C • Initial surface roughness: 1.4 to1.6 (μm)	• Surface roughness • Material removal

Table 3.1 Continued ……

Researchers (Year)	*Finished component (their material)*	*Process*	*Composition of the AFF medium*	*Used process parameters and their values*	*Responses*
Venkatesh et al. 2015	Straight bevel gear (EN-8 steel)	UA-AFF	Neoprene rubber, SiC abrasive particles	• Finishing time: 5, 10, 15 (min) • Abrasive particle size: 69 (μm) • Viscosity of the AFF medium: 0.73 (kPa.s) • Concentration of abrasive particles: 60 (wt %) • Flow rate of the AFF medium: 560 (cm^3/min) • Ultrasonic vibration frequency: 19 (kHz) • Ultrasonic vibration amplitude: 10 (μm) • Extrusion pressure: 30 (MPa)	• Surface roughness • Material removal

particles as the AFF medium and varying the finishing time. They also compared the finishing performance of the UA-AFF process with the AFF process and reported that (i) ultrasonic vibrations push the abrasive particles towards the workpiece bevel gear flank surfaces with a high velocity, which causes more abrasion to take place thus reducing its surface roughness and finishing time; (ii) surface finish and material removal increase with an increase in the finishing time; (iii) the surface finish of the workpiece bevel gear improves, having a glazed texture; (iv) the UA-AFF process reduced the same value of surface roughness in less finishing time than the AFF process; and (v) the improvement rate in surface roughness is 73% by the UA-AFF process and 55% by the AFF process.

3.2 Past Work on Laser Texturing

Laser texturing has recently emerged as an effective techno-commercial solution to improve the performance of various machining and finishing processes and enhance the tribological performance and service life of those engineering components subjected to fatigue loading, friction, wear, and higher operating temperatures, such as gear, camshaft, and cutting tools. The presence of the texture improves their friction and wear behavior and lubricating properties. Considerable research has been reported in the last three years on laser texturing. The following paragraphs describe them briefly, and table 3.2 presents the details not covered in the following paragraphs.

Xing et al. (2013) created laser texture on Si_3N_4 and TiC ceramics and investigated their anti-wear performance using a ball-on-disk type wear test and finite element analysis to study the stress distribution. They concluded that the (i) tribological performance is influenced by the size and density of the grooves created by the laser, (ii) wavy grooves with large density result in a lower coefficient of friction, and (iii) texturing improves the stress distribution pattern at the contact edges and reduces stress concentration. Sasi et al. (2017) performed laser texturing on the cutting tool made of high-speed steel (HSS) for the machining Al7075-T6 aluminum alloy for aerospace applications and found that it enhanced the tribological properties of the HSS cutting tool under dry machining conditions and reduced the cutting force and thrust force by 9% and 19%, respectively. Kang et al. (2017) laser textured an injection cam made of AISI 1045 steel to improve its anti-wear characteristics for its use in an internal combustion engine. They reported a 30% improvement in the anti-wear characteristics of the laser-textured injection cam compared to the non-textured injection cam. Niketh and Samuel (2017) created micro-texture and micro-dimples by laser on an 8 mm diameter carbide drill to make holes in Ti-6Al-4V titanium alloy, and reported a 12.3% reduction in torque, a 10.6% reduction in the thrust force, and formation of a less built-up edge (BUE) than that of a non-textured drill. Ye et al. (2018) created micro-grooved texture on the rack face of a cemented carbide cutting tool for turning C45 steel, and reported a significant reduction in the cutting forces and coefficient of friction compared to the non-textured tool. Hao et al. (2018) generated homothetic and hybrid textures by laser on carbide cutting tools for machining of Ti-6Al-4V titanium alloy, and reported a 9.3% reduction in the coefficient of friction for hybrid texture and 5.8% for homothetic texture compared to the non-textured tool. Singh et al. (2018) used laser to generate micro-patterns and micro-dimples on cellulose acetate film, polyethylene terephthalate, Ti-6Al-4V titanium alloy, and stainless steel (SS 304) to improve their tribological behavior and contact angle. They found that laser texturing increased the surface roughness, which increased the contact angle in both metal and polymers, and that the coefficient of friction reduced from 0.66 to 0.45, 0.18, and 0.17, respectively, for low, medium, and high-density areas of texture.

Melting and ablation of the workpiece material are observed in the laser texturing process due to heat generation. This necessitates subsequent finishing of the laser-textured component to remove the slag and excess molten material to achieve the intended surface quality. This could be achieved by the AFF process.

Table 3.2: Summary of past work on laser texturing of different materials

Researchers (Year)	*Type of laser used*	*Type of texture*	*Materials*	*Applications*	*Laser texturing parameters*	*Responses*
Xing et al. (2013)	• Pulsed Nd:YAG laser	• Homothetic • Wavy groove	• Silicon nitride • Titanium carbide	• Not mentioned	• Wavelength: 1064 (nm) • Pulse duration: 10 (ns) • Voltage: 19.5 (V) • Frequency: 6 (kHz) • Scanning speed: 5 (mm/s)	• Coefficient of friction
Sasi et al. (2017)	• Pulsed Nd:YAG laser	• Micro-dimples	• High-speed steel (HSS)	• Machining of aerospace alloy • (Al7075-T6)	• Wavelength: 355, 532, 1064 (nm) • Fluence: 40, 68, 100 (mW/cm^2) • Exposure time: 0.83, 1.16, 1.5 (min)	• Cutting force • Thrust force • Tool-chip contact length

Table 3.2 Continued ……

Researchers (Year)	*Type of laser used*	*Type of texture*	*Materials*	*Applications*	*Laser texturing parameters*	*Responses*
Kang et al. (2017)	• Pulsed Nd:YAG laser	• Micro-dimples	• Cast iron (AISI 1045 steel)	• Injection cam of engine	• Repetition frequen 20 (kHz) • Max. average pow 10 (W) • M^2 factor: ≤ 2 • Beam diameter: 3 (mm) • Laser beam expanders: 3 times • Focal length: 60 (n • Spot diameter: 60 (μm)	• Frictional force
Niketh and Samuel (2017)	• Pulsed Nd:YAG laser	• Micro-dimples • Homothetic	• Carbide drill	• Drilling in titanium alloy (Ti-6Al-4V)	• Wavelength: 1064 (nm) • Repetition frequency: 2.5 (kHz) • Pulse duration: 3.33×10^{-6} (min)	• Coefficient of friction • Thrust force • Torque • Chip morphology • Surface integrity

Table 3.2 Continued …

Researchers (Year)	*Type of Laser used*	*Type of texture*	*Materials*	*Applications*	*Laser texturing parameters*	*Responses*
Ye et al. (2018)	• Yb: KGW laser	• Homothetic	• Carbide tool	•Machining of C45 steel	• Max. average power: 15 (W) • Wavelength: 1030 (nm) • Pulse duration: 4.25 x10^{-9} (min) • Repetition frequency: 1 to 1.1 (MHz)	• Cutting force • Coefficient of friction
Hao et al. (2018)	• Pulsed fiber laser	• Hybrid • Homothetic	• Carbide tool	•Machining of titanium alloy (Ti-6Al-4V)	• Type of texture: Homothetic texture, Variable density texture, Variable shape texture, Variable shape and density texture • Mean density of texture: 40 % • Depth: 16.03(μm) • Width: 37.045 (μm)	• Coefficient of friction

Table 3.2 Continued …

Researchers (Year)	*Type of Laser used*	*Type of texture*	*Materials*	*Applications*	*Laser texturing parameters*	*Responses*
Singh et *al.* (2018)	• CO_2 laser	• Micro-pillar • Micro-dimples	• Acetate film • Polyethylene terephthalate • Titanium alloy • Stainless steel	• Not mentioned	• Power: 10; 14; 15; 18; 21 (W) • Frequency: 2.5; 5; 7 (KHz) • Scanning speed:1; 24; 60; 80 (mm/s) • Dots per inch: 400; 600; 1200 (dpi)	• Coefficient of friction

3.3 Past Work on Reduction of Noise and Vibrations of Gears

Researchers have used different traditional finishing processes to reduce the noise and vibrations of different types of gears by reducing their microgeometry errors and surface roughness. The following paragraph briefly describes them, and table 3.3 presents more details not covered in the following paragraph.

Yuruzume et al. (1979) used grinding to finish spur gears and studied the effects of improvement in their tooth profile on noise generation. They concluded that (i) an increase in speed and load on gear drive increases the noise in it, and (ii) noise generation in the gear drive is significantly influenced by profile error. Masuda et al. (1986) proposed a semi-empirical equation that has a dynamics term to represent the overall noise level generated by a gear. This equation considers the gear error characteristics in the vibration analysis of gear pairs, therefore it is capable of computing the noise levels for different gear finishing processes. They confirmed predictions by this equation for Niles-type grinding and Maag-type grinding of spur and helical gears made of Cr-Mo steel, and reported that the (i) gear noise level was reduced by 9.5 dB and 13.5 dB, respectively, for Niles-type and Maag-type grinding for both spur and helical gears, and (ii) the vibrations level was less in Maag-type grinding than Niles-type grinding for both spur and helical gears. Liu et al. (1990) conducted 242 experiments on spur gears used in the headstock of a machine tool to compare the noise of ground spur gears, and reported that (i) gear finishing has a significant impact on reducing the noise and vibrations of a gear, and a gear with a better surface finish produced less noise than an unfinished gear, (ii) a gear with small errors in its microgeometry and functional performance parameters has less average noise level, (iii) maximum and minimum values of noise were 85 dB and 75.7 dB, respectively, and (iv) gear noise level decreased by 5-6 dB through internal honing. Åkerblom and Pärssinen (2002) experimented on eleven different test gear pairs finished by gear shaving and gear grinding processes, and concluded that (i) the reduction of transmission error is an important excitation mechanism to reduce gear noise, (ii) improving surface finish reduces the gear noise level by approximately1 to 2 dB at a low torque level, (iii) an increase in lead error increases the gear noise level by 1 to 3 dB, (iv) shaved gears were less noisy than ground gears, (v) gears ground with the threaded grinding wheel are less noisy than those ground by the profile grinding wheel, (vi) an increase in the face width decreases the gear noise level by approximately 5 dB, and (vii) an increase in flank crowning decreases the gear noise level by 1 to 3 dB. Bihr et al. (2009) reported on the effects of gear microgeometry errors on noise generation. They compared the noise levels of two gearboxes with different microgeometry and concluded that gearboxes with fewer microgeometry errors generated less noise. Jolivet et al. (2014) compared the noise of the gears finished by the grinding and power honing processes using multiscale analysis based on the continuous wavelet transform. They studied the effects of the flank surface topography on the gear noise and reported that (i) the surface roughness produced by the two processes are very close to each other, and (ii) ground gears produce less noise than the power-honed gears, the difference being 0.2 dB at the first harmonic and 0.3 dB at the second and third harmonic. Jolivet et al. (2016) studied the role of the gear finishing process in reducing the noise of a helical gear made of steel in dry and wet lubricating conditions. They compared the surface roughness values of the unfinished helical gear, a gear finished by grinding with worm meshing, and a gear finished by power honing with the internal meshing process. They concluded that (i) the surface roughness generated by the finishing process has a significant impact on the noise and vibrations of a gear. Ground gears generate fewer vibrations than the power-honed gears, and (ii) dry surface conditions contribute more to vibrations than wet surface conditions.

Table 3.3: Summary of past work on the reduction of gear noise and vibrations by traditional finishing processes

Researchers (Year)	*Types of gear used*	*Gear material*	*Gear specifications*	*Used process parameters and their values*	*Finishing method*
Yuruzume et al. (1979)	• Spur gear	• Not mentioned	• Module: 3 (mm) • Number of teeth: 40 (Driver) • Number of teeth: 67 (Driver) • Face width: 25 (mm) • Pressure angle: 20°	• Applied load: 200, 600, 1000 (N) • Rotational speed: 300, 660, 920, 160, 1600, 2000 (rpm) • Meshing frequency: 0.2, 0.44, 0.613, 0.773, 1.067, 1.333 (kHz) • Lubricant: SAE#90 oil • Lubrication rate: 20-30 drop/min • Backlash: 120 (μm)	• Gear grinding

Table 3.3 Continued ……

Researchers (Year)	*Types of gear used*	*Gear material*	*Gear specifications*	*Used process parameters and their values*	*Finishing method*
Masuda et al. (1986)	• Spur gear • Helical gear	• Cr-Mo steel (SCM445) • Cr-Mo steel (SCM440)	• Module: 4 (mm) • Pressure angle: 25° • Helix angle: 10 • Face width: 18.5, 140 (mm) • Number of teeth: 22 (Driver) • Number of teeth: 86 (Driven)	• Power: 50-300 kW with increment of 50 kW • Rotational speed: 600-2200 rpm with an increment of 200 rpm • Applied load: 98-1372 N with increment of 98 N	• Hobbed gear • Niles-type grinding • Maag-type grinding
Liu et al. (1990)	• Spur gears of lathe headstock	• Not mentioned	• Module: 2.5 (mm) • Number of teeth: 40 • Face width: 20 (mm)	• Rotational speed: 1000 (rpm) • Applied load: 14.70 (Nm)	• Gear grinding • Gear honing
Åkerblom and Pärssinen (2002)	• Spur and helical gear	• Case-hardened steel (V-2525-94)	• Pressure angle: 20° • Helix angle: 20° • Number of teeth: 49 (pinion) • Number of teeth: 55 (gear)	• Applied load: 10, 50, 140, 500, 1000 (Nm) • Rotational speed: 600-2000 rpm with an increment of 200 rpm	• Gear shaving • Profile wheel grinding • Threaded wheel grinding

Table 3.3 Continued ……

Researchers (Year)	*Types of gear used*	*Gear material*	*Gear specifications*	*Used process parameters and their values*	*Finishing method*
Jolivet et al. (2014)	• Helical gear	• Steel	• Module: 1.85 (mm) • Pressure angle: 20° • Helix angle: 25° • Active face width: 24 (mm) • Number of teeth: 14; 23 (Driver) • Number of teeth: 51; 49 (Driven)	• Applied load: 8 (Nm) • Rotational speed: 500 to 6850 rpm	• Gear grinding • Power honing
Jolivet et al. (2016)	• Helical gear	• Steel	• Module: 1.85 (mm) • Pressure angle: 20° • Helix angle: 25° • Active face width: 24 (mm) • Number of teeth: 23 (Driver) • Number of teeth:51 (Driven)	• Applied load: 8 (Nm) • Rotation speed:1500 (rpm)	• Gear grinding • Power honing

3.4 Existing Gaps

The following are the identified gaps from the review of the relevant past work presented in previous sections:

- Past work focused on using the AFF process to reduce only the surface roughness of helical and straight bevel gears made of EN8 steel.
- No work is reported on reducing the microgeometry errors of straight bevel gears and the maximum and average surface roughness values and microgeometry errors of spur gears by the AFF process.
- No attempt was made to finish the gears made of 20MnCr5 alloy by the AFF process, which is most commonly used alloy for the commercial manufacturing of spur and bevel gears.
- No work is available to identify the optimum parameters of the AFF process, i.e. extrusion pressure, abrasive particle size, concentration of abrasive particle, concentration of the oil, and finishing time.
- No work attempted to improve the surface morphology, microhardness, and wear characteristics of spur and straight bevel gears finished by the AFF process.
- No work is available on reducing errors in functional performance parameters and the noise and vibrations of spur and straight bevel gears by the AFF process.
- No work reported on using laser texturing to improve the performance and productivity of the AFF process for gear finishing by reducing microgeometry errors and surface roughness and subsequently improving their surface morphology, microhardness, and wear resistance.

3.5 Objectives and Methodology

The following are the objectives of bridging the existing gaps using the investigation methodology depicted in fig. 3.1:

- To develop a robust AFF machine and fixtures for finishing the spur and straight bevel gears made of 20MnCr5 alloy by the AFF process. The developed fixtures should sustain high extrusion pressure and guide the AFF medium to reduce surface roughness, errors in microgeometry, and functional performance parameters of the spur and straight bevel gears.

- To study the effect of the AFF process parameters on the reduction of microgeometry error and surface roughness of the spur and straight bevel gears and identify optimum parameters.
- To study the surface morphology, microhardness, and wear resistance of the spur and straight bevel gears using the identified optimum parameters.
- To engage in a multi-response optimization of the AFF process parameters to simultaneously optimize the considered responses of the spur gear and straight bevel gear and their experimental validation.
- To study the role of laser texturing in improving the finishing performance and productivity of the AFF process by a comparative study of laser-textured and untextured spur and straight bevel gears finished by the AFF process in terms of microgeometry errors, surface roughness, surface morphology, wear resistance, microhardness, and MRR.
- To study the error reduction in the functional performance parameters, and noise and vibrations of the spur and straight bevel gears by finishing them with the AFF process.

Stage-1: Experiments to Identify Feasible Ranges of Finishing Time and Extrusion Pressure

Seven Experiments to Identify Feasible Range of Finishing Time

Variable parameter: Finishing time (Minutes): 10; 15; 20; 25; 30; 35; 40

Fixed parameters:

Extrusion pressure (MPa): 5

Abrasive particle concentration (Vol. %): 30

Abrasive particle size (Mesh): 100 (diameter: 150 μm)

Silicone oil concentration (Vol. %): 10

Responses: 'R_{max}' and 'R_a'

Four Experiments to identify feasible range of extrusion pressure

Variable parameter: Extrusion pressure (MPa): '2.5; 5; 7.5; 10

Fixed parameters:

Finishing time (Minutes): 15

Abrasive particle concentration (Vol.%): 30

Abrasive particle size (Mesh): 100 (diameter: 150 μm)

Silicone oil concentration (Vol.%): 10

Responses: Surface quality through visual examination

Stage-2: Twenty experiments to Identify Optimum Values of Finishing Time and silicone oil concentration of (i.e. Medium Viscosity)

Variable parameters

Finishing time (Minutes): 10; 15; 20; 25; 30

Silicone oil concentration (Vol. %): 10; 15; 20; 25

Medium viscosity (kPa S): 135; 54; 8; 2)

Fixed parameters

Extrusion pressure (MPa): 5

Abrasive particle concentration (Vol.%): 30

Abrasive particle size (Mesh): 100 (diameter: 150 μm)

Responses (Percentage reduction in following parameters)*

For spur gear only

- Total profile error 'PRF_a' and Total lead error 'PRF_β',

For bevel gear only

- Single pitch error 'PRf_p' and Adjacent pitch error 'PRf_u'

For both spur and bevel gears

- Total pitch error 'PRF_p' and Radial runout 'PRF_r'
- Max. surface roughness 'PRR_{max}' Avg. surface roughness 'PRR_a'

Laser Texturing of Gears to Improve Performance of AFF Process

- ***Full factorial Seventy-Two*** experiments for spur and straight bevel gears each to identify optimum values of parameters of 1064 nm fibre laser for producing ***homothetic texture***

Laser power (W): 10; 15; 20; 25; Focal length (mm): 280; 285; 290

Numbers of passes: 1, 2, 3, 4, 5, and 6

- Experiments (3) for AFF of **laser textured** spur and straight bevel gears each by varying finishing time

Variable parameter: Finishing time (Minutes): 15; 20; 25 for spur gears and 20; 25; 30 for straight bevel gears

Fixed parameters

Silicone oil concentration (Vol.%): 10

Extrusion pressure (MPa):5

Abrasive particle concentration (Vol.%): 30

Abrasive particle size (Mesh): 100 (diameter: 150 μm)

Responses (Percentage reduction in following parameters)*

For spur gear only

- Total profile error 'PRF_a' and total lead error 'PRF_β'

For both spur and bevel gears

- Total pitch error 'PRF_p' Radial runout 'PRF_r'
- Max. surface roughness 'PRR_{max}' and avg. surface roughness 'PRR_a'

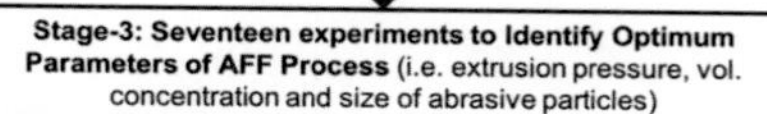

Stage-3: Seventeen experiments to Identify Optimum Parameters of AFF Process (i.e. extrusion pressure, vol. concentration and size of abrasive particles)

Variable parameters

Extrusion pressure (MPa): 3; 5; 7

Abrasive particle concentration (Vol. %): 20; 30;40

Abrasive particle size (Mesh): 80; 100; 120 (diameter: 190; 150; 127 μm)

Fixed parameters:

Finishing time (Minutes): 25 for Spur and 30 for Straight Bevel gear

Silicone oil concentration (Vol. %): 10

Responses (Percentage reduction in following parameters)*

For spur gear only

- Total profile error 'PRF_a' and total lead error 'PRF_β'

For both spur and bevel gears

- Total pitch error 'PRF_p' Radial runout 'PRF_r'
- Max. surface roughness 'PRR_{max}' and avg. surface roughness 'PRR_a'

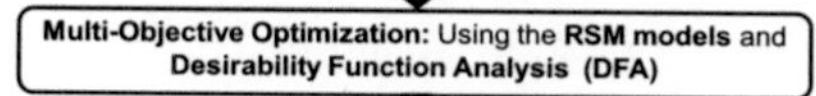

Multi-Objective Optimization: Using the **RSM models** and **Desirability Function Analysis (DFA)**

Validation of Optimization Results: By conducting **three confirmation experiments each** for spur gears and straight bevel gears

Dual Flank Roll Testing and Noise and Vibration Analysis: For the best finished spur and straight bevel gear using the optimized AFF parameters

* Surface morphology, Microhardness, Wear characteristics have been studied for the best finished spur and straight bevel gears
Finishing medium used in all the experiments: Mixture of SiC abrasive particles + silly putty + silicone oil

Fig. 3.1: Investigation methodology

References

[1] Åkerblom, M. and M. Pärssinen. 2002. "A study of gear noise and vibration," Trita-MMK (ISSN 1400-1179), 8, pp. 44. www.diva-portal.org/smash/get/ diva2:139881/FULLTEEXT01.pdf

[2] Bihr, J., M. Heider, M. Otto, K. Stahl, T. Kume and M. Kato. 2015. "Gear noise prediction in automotive transmissions." *Gear Technology: Journal of Gear Manufacturing*. 32(6): 66-70.

[3] Hao, X., X. Chen, S. Xiao, L. Li, and N. He. 2018. "Cutting performance of carbide tools with hybrid texture." *The International Journal of Advanced Manufacturing Technology*. 97(9): 3547-3556. doi: 10.1007/s00170-018-2188-2

[4] Jolivet, S., S. Mezghani, M. E. Mansori, and B. Jourdain. 2015. "Dependence of tooth flank finishing on powertrain gear noise." *Journal of Manufacturing Systems*. 37: 467-471. doi: 10.1016/j.jmsy.2014.11.006

[5] Jolivet, S., S. Mezghani, J. Isselin, and M. E. Mansori. 2016. "Experimental and numerical study of tooth finishing processes contribution to gear noise." *Tribology International*. 102: 436-443. doi: 10.1016/j.triboint.2016.06.005

[6] Kang, Z., Y. Fu, J. Ji and J. C. Puoza. 2017. "Effect of local laser surface texturing on tribological performance of injection cam." *The International Journal of Advanced Manufacturing Technology*. 92(5): 1751-1760. doi: 10.1007/s00170-017-0227-z

[7] Liu J., J. H. Wu., H. M. Qian., P. C. Chen. 1990. "Gear noise and the making of silent gears." *Gear Technology: Journal of Gear Manufacturing*. 7(2):8-15.

[8] Niketh, S. and G. L. Samuel. 2017. "Surface texturing for tribology enhancement and its application on drill tool for the sustainable machining of titanium alloy." *Journal of Cleaner Production*. 167: 253-270. doi: 10.1016/j.jclepro.2017.08.178

[9] Saeidi, F., M. Parlinska-Wojtan, P. Hoffmann and K. Wasmer. 2017. "Effects of laser surface texturing on the wear and failure mechanism of grey cast iron reciprocating against steel under starved lubrication conditions." *Wear*. 386-387: 29-38. doi: 10.1016/j.wear.2017.05.015

[10] Sasi, R., S. K. Subbu and I. A. Palani. 2017. "Performance of laser surface textured high-speed steel cutting tool in machining of Al7075-T6 aerospace alloy." *Surface and Coatings Technology*. 313: 337-346. doi: https://doi.org/10.1016/j.surfcoat.2017.01.118

[11] Singh, A., D. S. Patel, J. Ramkumar and K. Balani 2018. "Single-step laser surface texturing for enhancing contact angle and tribological properties." *The International Journal of Advanced Manufacturing Technology*. 100(5): 1253-1267. doi: 10.1007/s00170-018-1579-8

[12] Venkatesh, G., A. K. Sharma, N. Singh and P. Kumar. 2014. "Finishing of bevel gears using abrasive flow machining." *Procedia Engineering.* 97: 320-328. doi: 10.1016/j.proeng.2014.12.255

[13] Venkatesh, G., A. K. Sharma and P. Kumar. 2015. "On ultrasonic-assisted abrasive flow finishing of bevel gears." *International Journal of Machine Tools and Manufacture.* 89: 29-38. doi: 10.1016/j.ijmachtools.2014.10.014

[14] Xing, Y., J. Deng, X. Feng and S. Yu. 2013. "Effect of laser surface texturing on Si3N4/TiC ceramic sliding against steel under dry friction." *Materials & Design.* 52: 234-245. doi: 10.1016/j.matdes.2013.05.077

[15] Xu, Y. C., K. H. Zhang, S. Lu and Z. Q. Liu. 2013. "Experimental investigations into abrasive flow machining of helical gear." *Key Engineering Materials.* 546: 65-69. doi: 10.4028/www.scientific.net/KEM.546.65

[16] Ye, D., Y. Lijun, C. Bai, W. Xiaoli, W. Yang and X. Hui. 2018. "Investigations on femtosecond laser-modified microgroove-textured cemented carbide YT15 turning tool with promotion in cutting performance." *The International Journal of Advanced Manufacturing Technology.* 96(9): 4367-4379. doi: 10.1007/s00170-018-1906-0

[17] Yuruzume, I., H. Mizutani and T. Tsubuku. 1979. "Transmission errors and noise of spur gears having uneven tooth profile errors." *Journal of Mechanical Design.* 101(2): 268-273. doi: 10.1115/1.345404

CHAPTER FOUR

DEVELOPMENT OF THE AFF MACHINE AND FIXTURES

This chapter describes the development of the machine and the required fixtures to impart a high-quality finish to different types of gears by the two-way AFF process. The developed AFF machine should sustain extrusion pressure as high as 20 MPa.

4.1 Development of Machine for Two-way AFF Process

Figure 4.1 shows the vertical configuration machine for a two-way AFF process. It consists of two opposing hydraulic cylinders and two cylinders containing the viscoelastic medium used in the AFF process. The developed fixture containing the workpiece gear is located between the cylinders containing the AFF medium.

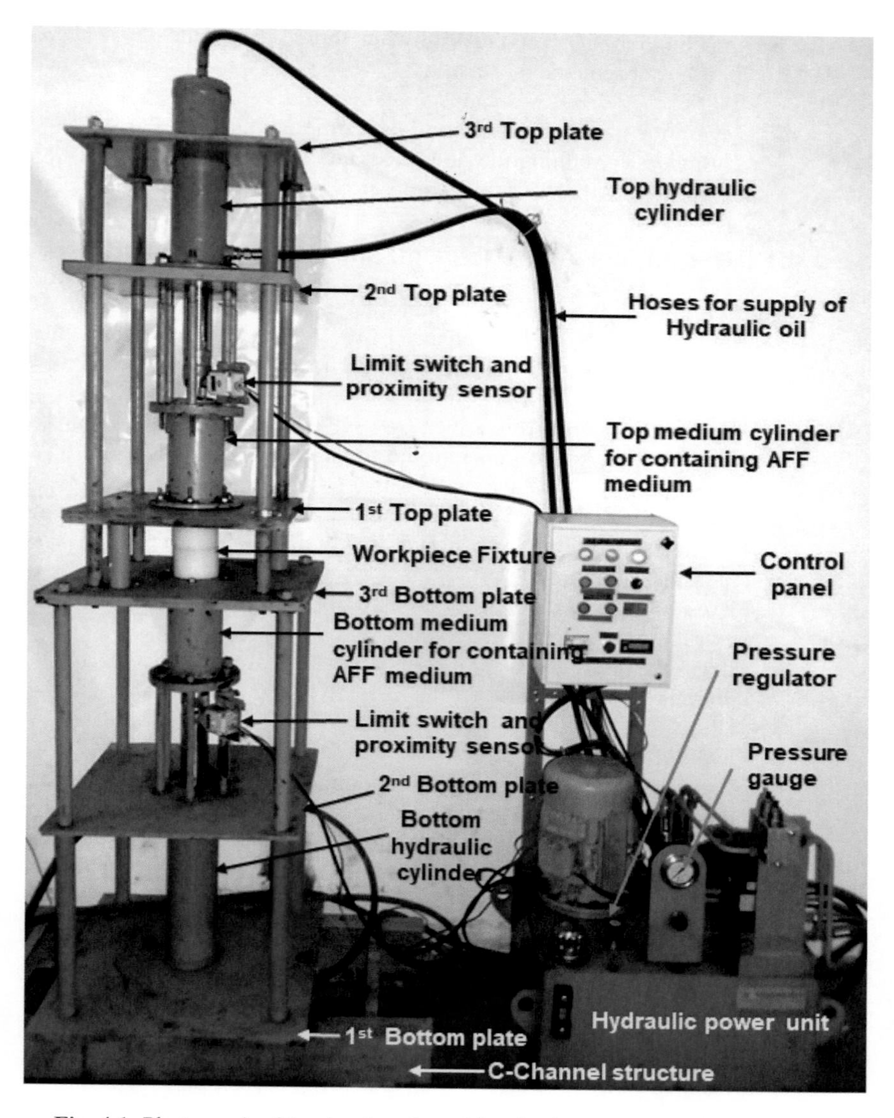

Fig. 4.1: Photograph of the developed machine for the two-way AFF process (Source: Petare and Jain, 2018, Reprinted with permission from Elsevier © 2018).

The machine consists of the following three subsystems, which are described in the subsequent subsections:

- Hydraulic power pack unit and hydraulic cylinders
- Cylinders containing the AFF medium
- Support structures

4.1.1 Hydraulic Power Pack Unit and Hydraulic Cylinders

The hydraulic power pack unit (manufacturer: Vaishnavi Hydraulics & Engineering Co., Indore, India) consists of the following components, the specifications of which are provided in table 4.1:

- **Hydraulic pump:** It is driven by 5 HP electric motor to pressurize the hydraulic oil to generate the required extrusion pressure to a maximum of 20 MPa.
- **Hydraulic cylinders:** Two hydraulic cylinders of 100 mm internal diameter and 300 mm stroke length are used to contain the pressurized hydraulic oil.
- **Pressure regulator:** It is used to set the desired value of the extrusion pressure.
- **Pressure gauge:** For indicating the set value of the extrusion pressure.
- **Limit switch** (Fig. 4.2a)**:** It controls the stroke length of a hydraulic cylinder.
- **Proximity sensor** (Fig. 4.2b)**:** It counts the number of strokes, i.e. movement of the hydraulic piston from top to bottom or from bottom to top is counted as one stroke, and two such strokes constitute one cycle. One set of a limit switch and proximity sensor is mounted in the space between the small threaded rods for the lower hydraulic cylinder, and the same is mounted for the upper hydraulic cylinder.
- **Control panel:** It monitors the flow direction of the hydraulic oil in the hydraulic cylinders. It has the option of setting the AFF process to be performed either in automatic or manual mode. It also has a display to indicate the completed number of strokes and a button to stop the AFF process in case of an emergency.

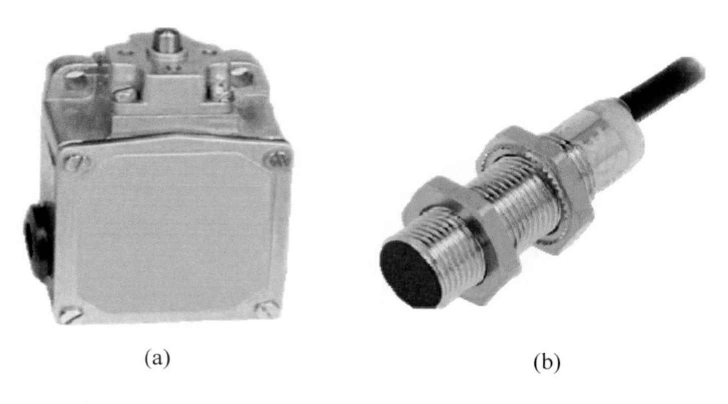

(a) (b)

Fig. 4.2: Photographs of (a) the limit switch used to control stroke length; and (b) the proximity sensor for counting number of strokes.

Table 4.1: Detailed specifications of the components of the hydraulic power pack unit

Name of component	Detail description	Value (unit)
Hydraulic pump	Max. extrusion pressure	20 (MPa)
	Power of electric motor	5 (HP)
	Oil flow rate	10 (liters per minute)
Hydraulic cylinders	Diameter of cylinder	100 (mm)
	Stroke length	300 (mm)
	Diameter of piston	63 (mm)
Limit switch	Current	10 (A)
	Voltage	500 (Volts)
Proximity sensor	Current	100-300 (mA)
	Voltage	5-40 (VDC)

4.1.2 Cylinders containing the AFF Medium

Two cylinders made of mild steel (MS) and with an 80 mm internal diameter and 250 mm stroke length were press-fitted with commercially available liners and the pistons to contain the AFF medium. The flanges, with six equispaced circumferential holes, were welded at the top and bottom of these cylinders to mount them between the supporting plates. Fig. 4.3 depicts a cylinder containing the AFF medium along with the liner, piston rings, and the piston used in it. The prepared AFF medium is poured into one of the medium-containing cylinders and its piston is moved back and forth by the hydraulic power pack unit, causing the back and forth extrusion of the AFF medium through the passage formed by the workpiece gear and its developed fixture.

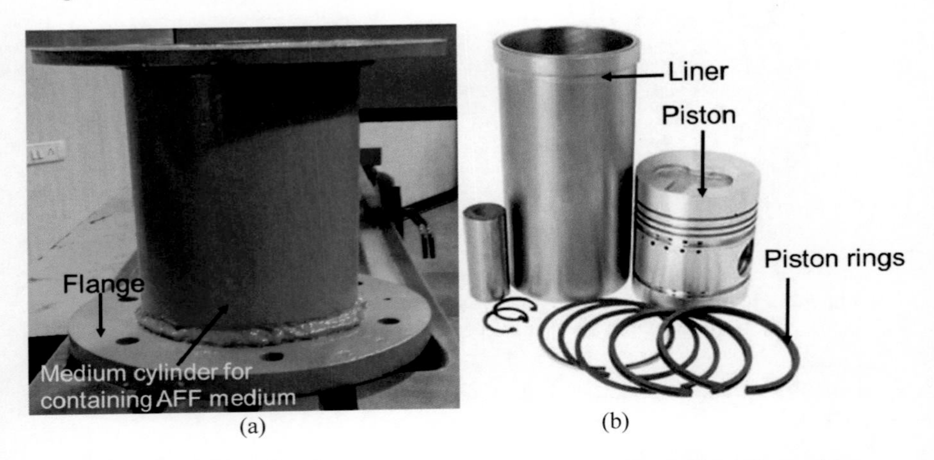

Fig. 4.3: Photograph of the (a) cylinder to contain the AFF medium along with flanges; and (b) cylinder liner and piston assembly.

4.1.3 Supporting Structure

A structure consisting of six MS plates (three plates, larger than the three plates fitted on the upper side, fitted on the lower side) and eight MS-stepped rods (four rods, larger than the four rods fitted on the upper side, fitted on the lower side) were manufactured to support and hold the hydraulic cylinders, the AFF medium-containing cylinders, and the developed fixture containing the workpiece gear. The heaviest plate was mounted as the bottom plate in the AFF machine on a channel structure and a lower hydraulic cylinder was mounted on it. The second lower plate was placed on the top face of the lower hydraulic cylinder. The lower cylinder containing the AFF medium was assembled on this plate with the support of six small threaded rods. The length of these rods was fixed according to the stroke length of the hydraulic cylinders. The third plate was fixed on this cylinder to provide a supporting base for mounting the developed fixture for the workpiece gear. A similar arrangement is repeated on the upper side of the workpiece fixture.

4.2 Development of Fixtures for Finishing the Gears

Special fixtures were developed for holding and supporting: (i) the spur gear (fig. 4.4a), (ii) helical gear (fig. 4.4b) by Rana et al. (2022), and (iii) the straight bevel gear (fig. 4.4c) to enable their finishing by the AFF process. Each fixture has two cylindrical discs made of Metlon, with a concentric hub made on a cylindrical protrusion for mounting the workpiece gear. The fixture for the spur gear (fig. 4.4a) has 16 circumferential holes of 5 mm in diameter, and the fixture for the helical gear (fig. 4.4b) has 21 circumferential holes of 5 mm in diameter drilled in both the discs at a radial location determined by their respective pitch circle diameter. The fixture for the straight bevel gear (fig. 4.4c) has 10 circumferential holes of 8 mm in diameter in upper disc and 5 mm in diameter holes in the lower disc drilled according to the outside and inside pitch circle diameters of the straight bevel gear, respectively. Four locating pins were provided at the circumference of the upper disc with 4 corresponding holes in the lower disc to avoid any relative displacement between them during the gear finishing by the AFF process. A central hole was drilled in both the discs to lock them together. The workpiece gear was mounted in its fixture in such a way that it is holds firmly against high extrusion pressure. The circumferential holes in the upper disc allow the entry of the AFF medium during the downward stroke, which is followed by finishing the flank surfaces of two consecutive teeth of the workpiece gear along their entire face width by the AFF medium and then the exit of the AFF medium from the circumferential holes in the lower disc. The sequence reverses in an upward stroke. This causes a shearing off of the surface peaks from the flank surfaces of the workpiece gear teeth, reducing errors in their microgeometry and surface roughness. Figure 4.5 schematically depicts the interaction of the abrasive particles contained in the AFF medium with the workpiece gear teeth for reducing its microgeometry errors.

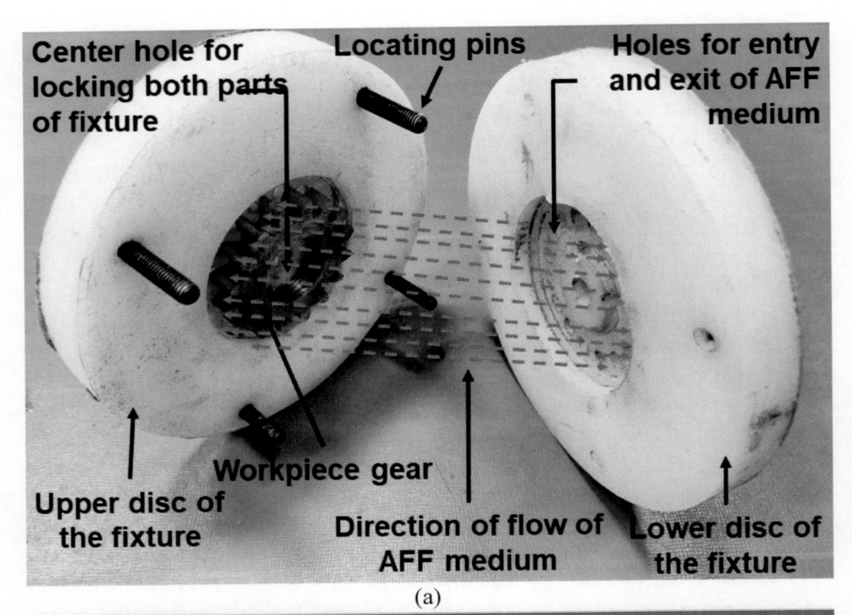
Center hole for locking both parts of fixture
Locating pins
Holes for entry and exit of AFF medium
Workpiece gear
Upper disc of the fixture
Direction of flow of AFF medium
Lower disc of the fixture

(a)

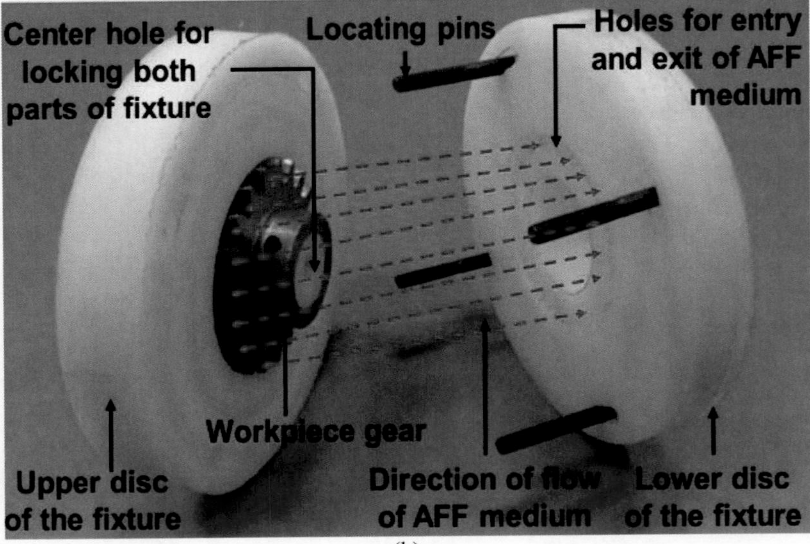
Center hole for locking both parts of fixture
Locating pins
Holes for entry and exit of AFF medium
Workpiece gear
Upper disc of the fixture
Direction of flow of AFF medium
Lower disc of the fixture

(b)

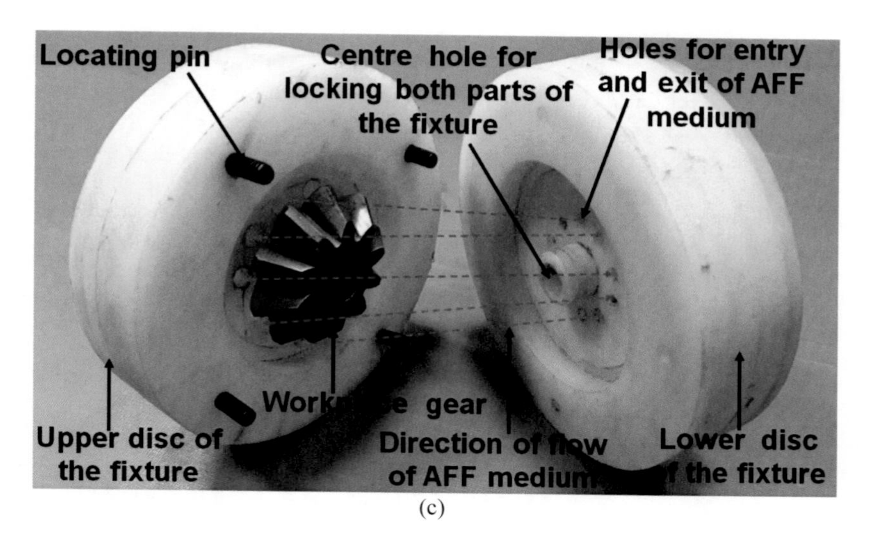

(c)

Fig. 4.4: Photograph of the developed fixture for holding and supporting the workpiece (a) spur gear; (b) helical gear (Rana et al., 2022), and (c) straight bevel gear, during their finishing by the AFF process (Petare et al., 2021)

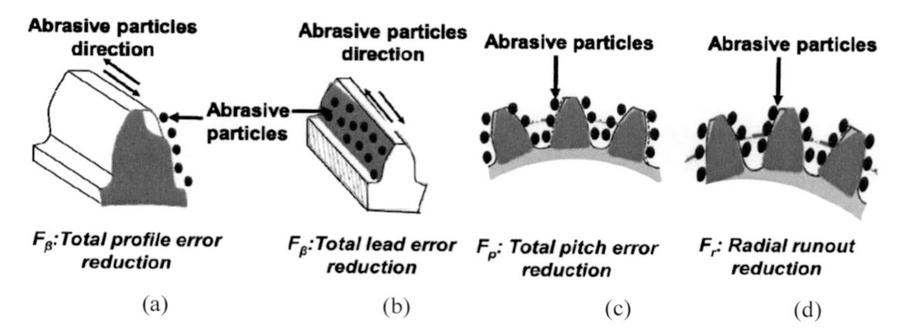

Fig. 4.5: Schematic of the abrasive particles of the AFF medium reducing gear microgeometry errors in: (a) total profile, (b) total lead, (c) total pitch, and (d) radial runout (Petare and Jain, 2018b)

4.3 Details of the Workpiece Gears

Alloy steel 20MnCr5, which is widely used for the commercial applications of spur and straight bevel gears [having 76.4 Rockwell hardness at B scale (HRB)], was selected for the in-house manufacturing of spur gears by the gear hobbing process, whereas straight bevel gears made of case hardened 20MnCr5 (having hardness 91.3 HRB) were procured. The chemical composition of 20MnCr5 (by wt.%) is 1.10% Cr, 1.19% Mn, 0.18% C, 0.019% P, and 0.017% S, and balance Fe (refer to Appendix-A for details). Figure 4.6 presents the specifications of the spur and straight bevel gears.

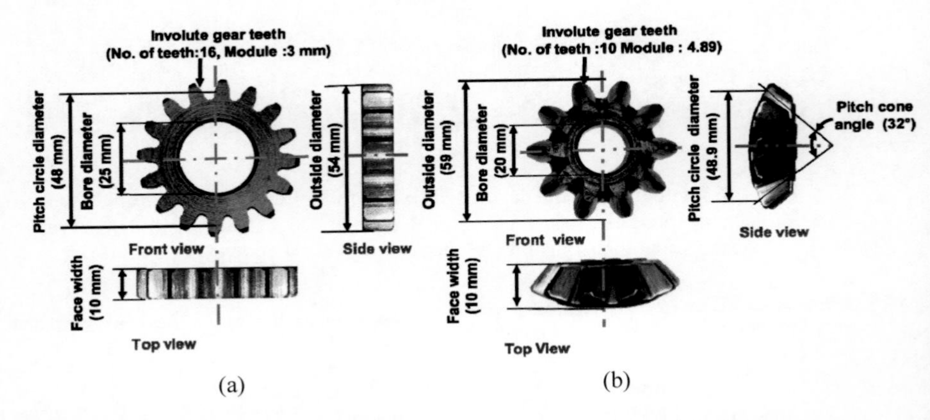

Fig. 4.6: Specifications of the workpiece (a) spur gear and (b) straight bevel gear (Petare et al., 2021)

References

[1] Petare, A. C. and N. K. Jain. 2018a. "On simultaneous improvement of wear characteristics, surface finish and microgeometry of straight bevel gears by abrasive flow finishing process." *Wear.* 404-405: 38-49. doi:10.1016/j.wear.2018.03.002

[2] Petare, A. C. and N. K. Jain. 2018b. "Improving spur gear microgeometry and surface finish by AFF process." *Materials and Manufacturing Processes.* 33(9): 923-934. doi:10.1080/10426914.2017.1376074

[3] Petare, A., N. K. Jain and I. A. Palani. 2021. "Simultaneous improvement of microgeometry and surface quality of spur and

straight bevel gears by abrasive flow finishing process." *Journal of Micromanufacturing.* 4(2): 189-206.
doi:10.1177/25165984211021010

[4] Rana, V., Anand Petare, Neelesh Kumar Jain, Anand Parey. 2022. "Using abrasive flow finishing process to reduce noise and vibrations of cylindrical and conical gears." *Proceedings IMechE, Part B: Journal of Engineering Manufacture.*
doi: 10.1177/09544054221075875

CHAPTER FIVE

EXPERIMENTATION DETAILS

The experimental investigation was planned, designed, and performed in the following stages:

- **Stage-1:** Experiments were conducted in two phases to identify the feasible ranges of the finishing time and extrusion pressure.
- **Stage-2:** Full factorial experiments were conducted to identify the optimum values of the finishing time and viscosity of the AFF medium.
- **Stage-3**: Experiments were conducted to identify the optimum parameters of the AFF process using the Box-Behnken (BBD) experimental design approach of the response surface methodology (RSM).
- **Stage-4:** Experiments to validate the results of optimization by the desirability functional analysis (DFA).

This chapter provides details of the preparation of the AFF medium, experimentations in each stage along with the used approach for designing the experiments, details of the considered responses and their measurement, details of laser texturing, and the performance characteristics (determined by the dual flank roll testing, and noise and vibrations testing) of the unfinished and the best finished gears by the AFF process.

5.1 Preparation of the AFF Medium and its Viscosity Measurement

The AFF medium consists of the abrasive particles chosen according to type (i.e. ductile or brittle, metal, alloy, or composite) and hardness of the workpiece material, long-chain polymer-based putty, and a blending oil (Gorana et al., 2006). Silicon carbide was selected as the abrasive, keeping in view the hardness of the selected gear material. The recommended size range for silicon carbide abrasive particles for the gear grinding process (for gear module range 1.5-5 mm) is 80 to 120 mesh, i.e. the corresponding diameter of the abrasive particles 'd_a' is 190 to 127 μm (Norton Technical Guide, 2017). Consequently, the 80 to 120 mesh size of abrasive particles was chosen. The range of volumetric concentration of the abrasive particles was chosen as 20 to 40% based upon the suggestion by Jain et al. (2007) and Bremerstein et al. (2015). Molding clay was chosen as a putty material due to its good ability to hold the AFF medium together, low cost, and easy availability. Silicone oil was used as the blending oil because it enables the viscosity of the AFF medium to be controlled easily. A volume of 1156 cm^3 of the AFF medium (determined from the dimensions of the cylinder containing the AFF medium) of the selected abrasive particles, putty, and blending oil (whose quantities are calculated according to the required composition of the AFF medium) was thoroughly hand mixed then pressed in a deep drawing machine. The viscosity of the AFF medium was changed by increasing the volumetric concentration of the silicone oil, which correspondingly reduces the volumetric percentage of the putty in the AFF medium of 1156 cm^3 volume. The viscosity of the AFF medium was measured using the rheometer MCR-301 by Anton Paar, Germany (please refer to Appendix-B for its details).

5.2 Procedure of the Experimentation

The following procedure was used to conduct the experiments at the different stages:

- All the considered responses were measured for all the unfinished spur gears and the unfinished straight bevel gears used in the experiments.
- The AFF medium of the required composition and concentration was prepared and poured into the lower cylinder. The extrusion pressure was fixed at the required value by the pressure regulator.
- A dry run was conducted prior to conducting the experiments. In this, the developed fixture for the spur gear or straight bevel gear was placed between the two cylinders containing the AFF medium, and the AFF medium was extruded back and forth for 10-15 minutes to ensure the proper mixing of its ingredients.
- For experimentation on the spur gears, each drilled hole in the lower disc of its developed fixture was aligned with the spacing between two adjacent teeth of the corresponding gear and with the drilled hole in the upper disc in such a way that the AFF medium in the upward stroke enters from a hole in lower disc, finishes flanks of two adjacent gear teeth, and exits from a corresponding

hole in the upper disc. This sequence of events reverses in the downward stroke. The same procedure was followed for the experiments on the straight bevel gears.

- The values of each input parameter were selected according to the plan of the experimentation at each stage.
- The finishing time for each experiment was recorded using a stopwatch, and the AFF process was stopped immediately after completion of the selected finishing time.
- Each workpiece gear after the AFF process was cleaned using turpentine oil and dipped into a lubricating oil to prevent rusting.
- All the considered responses were measured for all the finished spur and straight bevel gears by the AFF process.
- Since all the considered responses are of the smaller-the-better type, the higher reduction in any considered response is desirable, and its limiting value is 100%, i.e. the value of that response after finishing by the AFF process should be zero.

5.3 Stage-1: Identification of Feasible Ranges of Finishing Time and Extrusion Pressure

Experiments at stage-1 were conducted in two phases. Seven experiments for spur gears and seven experiments for straight bevel gears were conducted in phase-1 to identify a feasible range of finishing time 't' by varying it at seven levels (i.e. 10, 15, 20, 25, 30, 35, and 40 minutes) and using constant values of the extrusion pressure 'P' as 5 MPa, the size of abrasive particles 'M_a' as 100 mesh (with a corresponding diameter of 150 μm), the volumetric concentration of the abrasive particle 'C_{av}' as 30%, and the volumetric concentration of silicone oil 'O_c' as 10%. In phase-2, four experiments were conducted for spur and straight bevel gears each to identify a feasible range of the extrusion pressure by varying it at four levels (i.e. 2.5, 5, 7.5, and 10 MPa) and using constant values of finishing time 't' as 15 minutes, the size of abrasive particles as 100 mesh (with a corresponding diameter of 150 μm), the volumetric concentration of abrasive particles as 30%, and the volumetric concentration of silicone oil as 10%. Changes in maximum surface roughness 'R_{max}' and average surface roughness 'R_a' were used as the responses in both phases of the experimentation.

5.4 Stage-2: Identification of Optimum Values of Finishing Time and AFF Medium Viscosity

Twenty full-factorial experiments for the spur gears and 20 full-factorial experiments for the straight bevel gears were conducted in stage-2 to identify the optimum values of finishing time and volumetric concentration of silicone oil (used as a measure of viscosity of the AFF medium) by varying them at five levels (i.e. 10, 15, 20, 25, and 30 minutes) and at four levels (i.e. 10%, 15%, 20%, and 25% with corresponding values of viscosity of the AFF medium being 135, 54, 8, and 2 kPa. S, respectively. The values of extrusion pressure (5 MPa), size of abrasive particles (100 mesh), and volumetric concentration of abrasive particles (30%) were kept constant. The following were the objectives of these experiments:

- To study the effect of the variation in the finishing time and volumetric concentration of the silicone oil to identify their optimum values, which will simultaneously maximize the improvement in the considered microgeometry errors (i.e. total profile error 'F_a' and total lead error 'F_β' for the spur gear only; single pitch error 'f_a' and adjacent pitch error 'f_u' for the straight bevel gear only; total pitch error 'F_p', radial runout 'F_r' for both the spur and straight bevel gears), and maximum surface roughness 'R_{max}' and average surface roughness 'R_a' (for both the spur and straight bevel gears). A gear finished using the identified optimum values, which maximizes the reductions in its microgeometry errors (i.e. minimizes its microgeometry errors), is referred to as the best finished gear in stage-2.
- To study surface morphology, wear characteristics, and microhardness of the unfinished and the best finished spur and straight bevel gears.

The stage-2 experimentation identified the optimum values of the finishing time as 25 minutes for the spur gear and 30 minutes for the straight bevel gears and the volumetric concentration of silicone oil as 10% for use in stage-3 experiments.

5.5 Stage-3: Experimental Identification of Optimum Parameters of the AFF Process

The full factorial design of the experimental approach does not allow an evaluation of the interactions between the process parameters, therefore the Box-Behnken design (BBD) approach of the RSM was found to be suitable for designing the fractional factorial experiments of stage-3 because it provides more complete information in a minimum number of experiments compared to other fractional factorial approaches. The BBD approach was proposed by G. E. P. Box and Donald Behnken in 1960 and is represented graphically in fig. 5.1 for a three-factor, three-level experiment. It is a type of rotatable or nearly rotatable 2nd-order fractional factorial design in which each input parameter varies at three levels, usually coded as -1, 0, +1. The parametric combination for each experiment is at the midpoints of the edges of the process parameters space and the center. The total number of required experiments in the BBD approach is equal to $2k(k-1) + C_o$; where k is the number of input parameters and C_o is the number of central points, whereas the total number of experiments required using the central composite design (CCD) is given by $2^k + 2k + C_o$. The BBD approach has a more limited capability for orthogonal blocking than the CCD approach. Hence, if there is a requirement to separate the experimental runs into blocks for the BBD, then it allows the blocks to be used in such a manner that the calculations of the regression for the variable effects are not affected by these blocks.

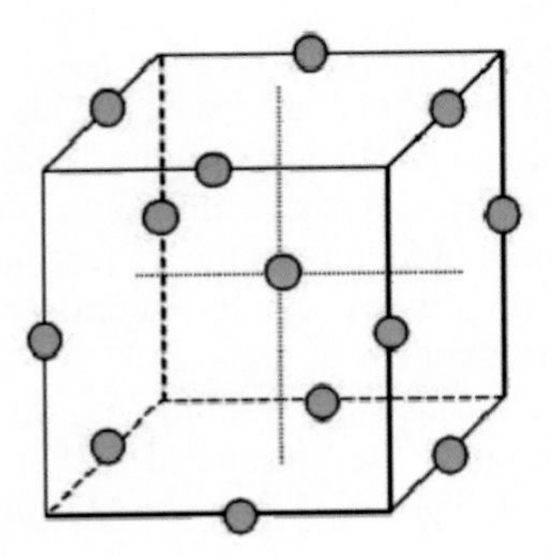

Fig. 5.1: Representation of the BBD of RSM for three parameters varying at three levels.

Accordingly, 17 experiments (including 5 experiments for central points) were designed and conducted for the spur and straight bevel gears each by varying the three most-important parameters at three levels each, i.e. extrusion pressure at 3, 5, and 7 MPa, abrasive particles at 80, 100, and 120 mesh (with corresponding diameters of 190, 150, and 127 μm), and volumetric concentrations of abrasive particles at 20, 30, and 40%. Table

5.1 presents the actual and coded values of these parameters for 17 experimental runs of stage-3, whose objectives are the following:

- To study the influence of the extrusion pressure, the size of abrasive particles, and the volumetric concentration of abrasive particles on the considered responses of microgeometry errors and surface roughness of spur gears and straight bevel gears.
- To identify the order of significance of the considered variable parameters and their interactions on the considered responses for spur gears and straight bevel gears.
- To identify the optimum parametric combinations that will simultaneously maximize the percentage reductions in the considered responses (microgeometry errors, i.e. total profile error, total lead error, total pitch error, and radial runout for the spur gears, and only total pitch error and radial runout for the straight bevel gears, and maximum and average surface roughness values for both gears). A gear finished using such a parametric combination is referred to as the best finished gear in stage-3.
- To study the surface morphology, wear characteristics, and microhardness of the unfinished and best finished spur gear and straight bevel gear.

Table 5.1: Actual and coded values of extrusion pressure 'P', abrasive particle size 'M_a', and volumetric concentration of abrasive particles 'C_{av}' for 17 experimental runs for Stage-3 experiments

Exp. run	Process parameters and symbol (unit)					
	Extrusion pressure 'P' (MPa)		*Abrasive particle size* 'M_a' (Mesh)		*Volumetric concentration of abrasive particle* 'C_{av}' (%)	
	Actual	*Coded*	*Actual*	*Coded*	*Actual*	*Coded*
1	7	(1)	120	(1)	30	(0)
2	7	(1)	80	(-1)	30	(0)
3	5	(0)	120	(1)	20	(-1)
4	5	(0)	100	(0)	30	(0)
5	5	(0)	100	(0)	30	(0)
6	7	(1)	100	(0)	20	(-1)
7	5	(0)	80	(-1)	20	(-1)
8	5	(0)	100	(0)	30	(0)
9	5	(0)	100	(0)	30	(0)
10	3	(-1)	80	(-1)	30	(0)
11	3	(-1)	100	(0)	40	(1)
12	5	(0)	100	(0)	30	(0)

13	7	(1)	100	(0)	40	(1)
14	5	(0)	120	(1)	40	(1)
15	3	(-1)	120	(1)	30	(0)
16	5	(0)	80	(-1)	40	(1)
17	3	(-1)	100	(0)	20	(-1)

5.6 Stage-4: Experimental Validation of the Optimization Results

Three experiments were conducted on spur and straight bevel gears each to validate the results of the parametric optimization by the desirability function analysis (DFA), with the objectives being the considered response. The DFA-optimized values of extrusion pressure, size of abrasive particles, and volume concentration of abrasive particles were compared with the optimum parametric combination identified by the RSM from the stage-3 experiments for the spur and straight bevel gears.

5.7 Evaluation of the Responses

Only surface roughness parameters (i.e. maximum surface roughness 'R_{max}', and average surface roughness 'R_a' for both the spur and straight bevel gears) are used as the responses for the stage-1 experimentation. Microgeometry errors (i.e. total profile error 'F_a' and total lead error 'F_β' for the spur gear only, single pitch error 'f_a' and adjacent pitch error 'f_u' for the straight bevel gear only, and total pitch error 'F_p' and radial runout 'F_r' for both the spur and straight bevel gears) and the surface roughness parameters were used as the responses for the experiments in stages 2, 3, and 4. Microgeometry errors, surface roughness parameters, and material removal rate (MRR) are used as responses to compare the finishing performance of laser-textured and untextured gears by the AFF process. Surface morphology, wear characteristics, and microhardness are used to study the best finished spur and straight bevel gears by the AFF process. Functional performance parameters determined from the dual flank roll testing and the changes in the noise and vibration levels are used to evaluate the performance characteristics of the unfinished and the best finished gears by the AFF process. The following subsections describe the evaluation of the various responses.

5.7.1 Measurements of Microgeometry Errors

Microgeometry errors of the unfinished and finished spur and straight bevel gears were measured on the computer numerical controlled (CNC) gear metrology machine SmartGear 500 (from Wenzel GearTec, Germany, please refer to Appendix-B for its details) by using a 3 mm-diameter probe. Values of total profile error 'F_a' and total lead error 'F_β' for the unfinished (i.e. before AFF) and finished (i.e. after AFF) spur gears were measured on the left-hand (LH) and right-hand (RH) flank surfaces of four randomly chosen teeth, and an average of the measured eight values was taken for further analysis. Single pitch error 'f_p' and adjacent pitch error 'f_u' (for bevel gear only), and total pitch error 'F_p' and radial runout 'F_r' for both the unfinished and finished spur and straight bevel gears were measured on the LH and RH flank surfaces of all the teeth, and their arithmetic mean was used during the further study. The arithmetic means of the measured values of total profile error before and after finishing by the AFF process were used to compute percentage reduction in it (i.e. Avg. PRFa) using Eq. 5.1.

$$Avg.PRF_a = \frac{Avg.\ F_a\ of\ a\ gear\ before\ AFF - Avg.\ F_a\ of\ the\ same\ gear\ after\ AFF}{Avg.F_a\ of\ the\ gear\ before\ AFF} 100(\%) \quad (5.1)$$

Similarly, the average values of percentage reductions in total lead error 'PRF_β', single pitch error 'PRf_p', adjacent pitch error 'PRf_u', and total pitch errors 'PRF_p' were calculated. Values of radial runout of the spur and straight bevel gears before and after finishing by the AFF process were used to calculate the percentage reduction in the radial runout 'PRF_r'.

5.7.2 Measurements of Surface Roughness Parameters

Maximum surface roughness 'R_{max}' and average surface roughness 'R_a' were measured by tracing 10 μm-diameter probe across the flank surface on two different locations on the LH and RH of two randomly chosen teeth of a spur gear and straight bevel gear (i.e. total eight values) before and after its finishing by the AFF process. The measurements were performed on the 3D surface roughness measuring cum contour tracing equipment *LD-130 MarSurf* (from *Mahr Metrology, Germany*, please refer to Appendix-B for its details). Evaluation lengths of 4 mm, cut-off length of 0.8 mm, and Gaussian filter were used to distinguish between roughness and waviness. The arithmetic means of all eight measured values of 'R_a' or 'R_{max}' were used to calculate the average percentage reduction in them, i.e. average percentage reduction in maximum surface roughness 'PRR_{max}' was calculated by Eq. 5.2.

$$Avg.PRR_{max} = \frac{Avg.\ R_{max}\ of\ a\ gear\ before\ AFF\ -\ Avg.\ R_{max}\ of\ the\ same\ gear\ after\ AFF}{Avg.\ R_{max}\ of\ a\ gear\ before\ AFF} 100\ (\%) \quad (5.2)$$

Similarly, the averagw percentage reduction in average surface roughness 'PRR_a' was calculated. Eq. 5.3 was used to calculate the change in the maximum surface roughness 'ΔR_{max}' used in the stage-1 experiments.

$$\Delta R_{max} = Avg.R_{max}\ of\ a\ gear\ before\ AFF\ -\ Avg.\ R_{max}\ of\ the\ same\ gear\ after\ AFF\ \ (\mu m) \quad (5.3)$$

Similarly, the change in average surface roughness 'ΔR_a' was calculated.

5.8 Evaluation of the Responses for the Best Finished Gears

The surface roughness profile, surface morphology, microhardness, and wear characteristics of the best finished spur and straight bevel gears during the stage-2 and stage-3 experiments were evaluated to get insight into the finishing mechanism, mechanical damage to the gear material, and improvement in their wear resistance after their finishing by the AFF process. The following paragraphs give brief details about their evaluation.

5.8.1 Assessment of Surface Integrity

Surface morphology, microhardness, and wear characteristics are used to assess the surface integrity of the gears after their finishing by the AFF process. Their evaluation methodology is briefly mentioned in the following subsections.

5.8.1.1 Surface Morphology Study

The surface morphology of the flank surfaces of the unfinished and the best finished spur and straight bevel gears were obtained by a field emission scanning electron microscopic (FESEM) *Supra 55* (from *Carl-Zeiss NTS GmbH, Germany*, please refer Appendix-B for its details) to assess the finishing mechanism by the AFF process.

5.8.1.2 Microhardness Evaluation

Values of the microhardness of the best finished and unfinished spur and straight bevel gears were measured on the Vickers microhardness tester *VMH-002* (from Walter UHL, Germany, please refer to Appendix-B for its details) by applying a load of 50, 100, and 200 (gm) for 15 seconds as per the standard ASTM E92-82 (2003). Three indentations were done for each load, and the average of these values was taken to study whether the AFF process alters the hardness of a gear material or not.

5.8.2 Study of Wear Characteristics

A fretting wear test was performed on one randomly selected tooth of an unfinished spur and an unfinished straight bevel gear, and one tooth of the best finished spur and best finished straight bevel gear on the tribometer *CM-9104* (from *Ducom, India,* please refer to Appendix-B for its details) as per the scheme depicted in figs. 5.2(a) and 5.2(b). A load of 50 N was applied over a 5 mm-diameter stainless-steel ball, which slides 5 mm at a frequency of 20 Hz for 20 minutes over the flank surface of the chosen spur or straight bevel gear tooth. The tribometer gives the variation of the friction force and coefficient of friction with time in the form of the graphs. Specific wear rate 'k_i' (mm^3/Nm), which is also known as Archard's coefficient (1953), was calculated using Eq. 5.4.

$$k_i = \frac{m_i}{\rho\, F\, S} \left(\frac{mm^3}{Nm}\right) \tag{5.4}$$

Where, 'F' is the applied normal load (N); 'S' is the total sliding distance, which is equal to two times the product of the sliding distance, frequency, and time duration (i.e. 240 m in the present case); 'm_i' is the mass loss during the wear test (mg); and 'ρ' is the density of the gear material (kg/mm^3). The weight loss of the gear tooth was measured using a precision weighing balance with an accuracy of 0.01 mg. Wear rate is a product of a specific wear rate and the applied load, and wear volume is the product of wear rate and the total sliding distance.

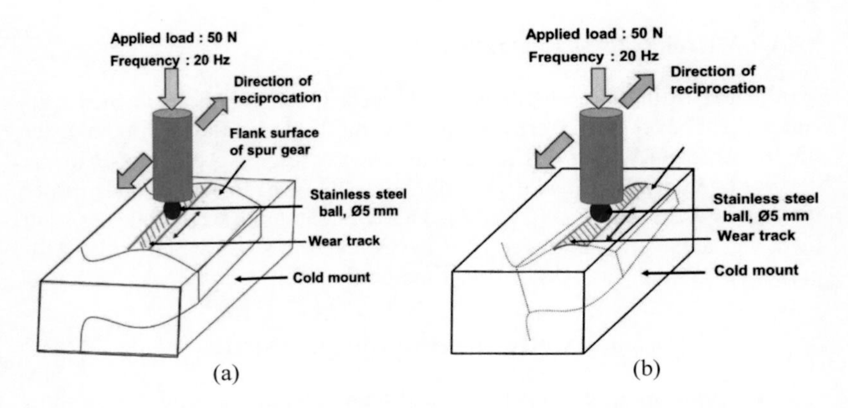

Fig. 5.2: Schematic of fretting wear test performed on the flank surface of randomly chosen tooth of the (a) spur gear and; (b) straight bevel gear.

5.8.3 Evaluation of Material Removal Rate

The average value of volumetric MRR for the AFF process is computed by Eq. 5.5, in which the average mass loss of workpiece gear during the AFF process is divided by the product of the finishing time and density of the gear material. The mass loss of a workpiece gear was calculated by taking the difference in its mass before and after the AFF process. The mass of a gear was measured on the precision balance *DS-852* (from *Essae Teraoka Ltd, India*, please refer to Appendix-B for its details), with an accuracy of 0.01 mg.

$$MRR = \frac{Mass\ of\ gear\ before\ AFF - Mass\ of\ the\ same\ gear\ after\ AFF}{Finishing\ time\ x\ Density\ of\ the\ gear\ material} \left(\frac{mm^3}{min}\right) \quad (5.5)$$

References

[1] Archard, J. F. 1953. "Contact and rubbing of flat surfaces." *Journal of Applied Physics.* 24(8): 981-988. doi:10.1063/1.1721448

[2] ASTM E92-82, 2003. Standard Test Method for Vickers Hardness of Metallic Materials, ASTM International, West Conshohocken, PA.

[3] ASTM G133-05, 2016. Standard Test Method for Linearly Reciprocating Ball-on-Flat Sliding Wear, ASTM International, West Conshohocken, PA.

[4] Bremerstein, T., A. Potthoff, A. Michaelis, C. Schmiedel, E. Uhlmann, B. Blug, and T. Amann. 2015. "Wear of abrasive media and its effect on abrasive flow machining results." *Wear.* 342-343: 44-51. doi:10.1016/j.wear.2015.08.013

[5] Gorana, V. K., V. K. Jain and G. K. Lal. 2004. "Experimental investigation into cutting forces and active grain density during abrasive flow machining." *International Journal of Machine Tools and Manufacture.* 44(2): 201-211.
doi: 10.1016/j.ijmachtools.2003.10.004

[6] Jain, N. K., V. K. Jain and S. Jha. 2007. "Parametric optimization of advanced fine-finishing processes." *The International Journal of Advanced Manufacturing Technology.* 34(11): 1191-1213.
doi:10.1007/s00170-006-0682-4

[7] Norton Technical Guide. 2017.Technical Solution for Grinding in the Gear Market, Saint- Gobain Abrasifs: Confalns Cedex, France, pp. 12-13.

CHAPTER SIX

EXPERIMENTAL FINDINGS AND DISCUSSION

This chapter presents the findings, their discussion, and conclusions drawn from each stage of the experimentation.

6.1 Findings from the Stage-1 Experiments

The stage-1 experimentation identified feasible ranges of finishing time and extrusion pressure, which are described in the following two subsections.

6.1.1 Identification of Feasible Range of Finishing Time

Table 6.1 presents the results of the seven experiments conducted in phase-1 of stage-1 to identify a feasible range of the finishing time for the spur and straight bevel gears finished by the AFF process by varying it seven levels. Figure 6.1 depicts the variation of changes in the maximum and average surface roughness values of the spur gear (fig. 6.1a) and straight bevel gear (fig. 6.1b), with the finishing time using the regression equation obtained using the results of table 6.1. It can be observed from the graphs of fig. 6.1 that:

- Changes in the maximum and average surface roughness values during the finishing time of 10-15 minutes are fewer due to the presence of higher surface roughness peaks, which the abrasive particles of the AFF medium try to reduce to attain approximately equal heights after 15 minutes of the finishing by the AFF process. Changes in surface roughness values increase rapidly during the 20 to 30 minutes of the finishing and attain their maximum values during 30 to 32 minutes of the finishing time for both the spur and straight bevel gear (i.e. their max. and avg. surface roughness values after AFF attain their minimum values in this range) due to the availability of surface roughness peaks of approximately equal heights.

- Changes in surface roughness values start decreasing for the finishing time beyond 35 minutes for both the spur and straight bevel gear because the cutting edges of the abrasive particles become blunt.

Therefore, a finishing time range of 10 to 30 minutes was identified as the feasible range for further experiments on the AFF of spur and straight bevel gears.

Table 6.1: Results of the Phase-1 experiment of Stage-1

Exp. run	**Finishing time 't' (minutes)**	**Responses**			
		Spur gear		**Straight bevel gear**	
		ΔR_{max} (μm)	**ΔR_a (μm)**	**ΔR_{max} (μm)**	**ΔR_a (μm)**
1	10	0.04	0.04	0.04	0.02
2	15	0.28	0.10	0.11	0.06
3	20	0.397	0.12	0.17	0.08
4	***25***	***0.87***	***0.22***	0.23	0.11
5	***30***	0.85	0.21	***0.39***	***0.15***
6	35	0.75	0.18	0.37	0.14
7	40	0.72	0.15	0.26	0.09

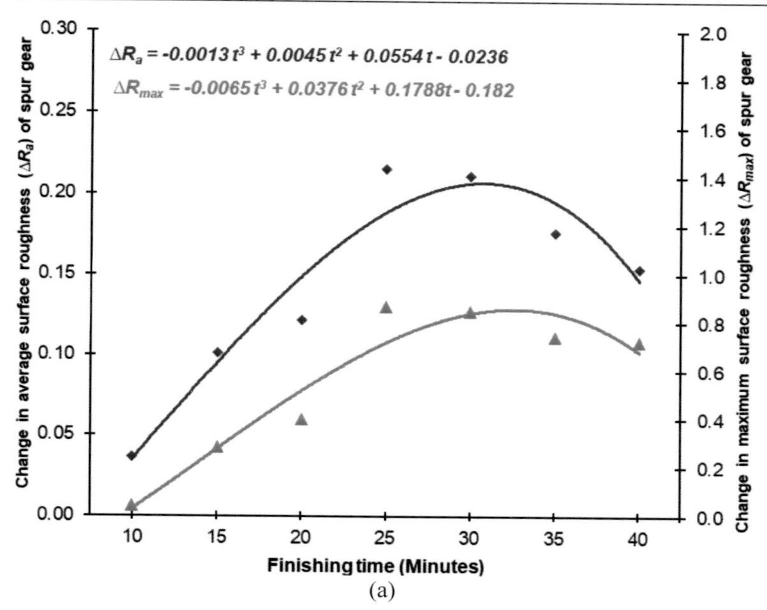

(a)

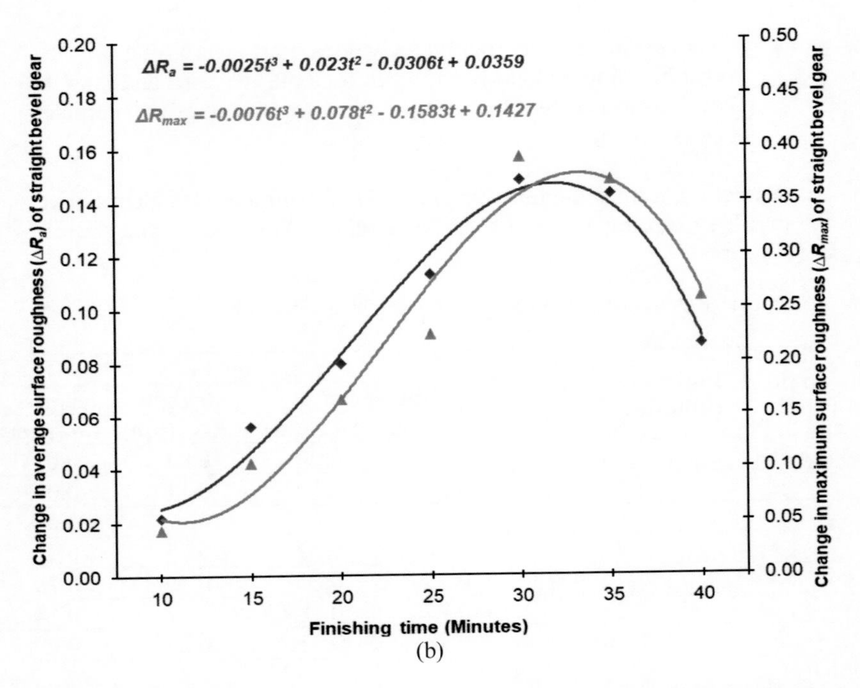

(b)

Fig. 6.1: Variation of change in maximum surface roughness and average surface roughness values with the finishing time for (a) spur gears; and (b) straight bevel gears (Petare et al., 2020; and Petare et al., 2021)

6.1.2 Identification of Feasible Range of Extrusion Pressure

Table 6.2 presents the results of the four experiments conducted in phase-2 of stage-1 to identify the feasible range of the extrusion pressure for both spur and straight bevel gears along with considered responses, i.e. changes in their maximum and average surface roughness values. It can be seen in table 6.2 that changes in the max. and avg. surface roughness values of the spur and straight bevel gears increase with extrusion pressures up to 7.5 MPa and start decreasing beyond this value. Moreover, it is observed that using the extrusion pressure beyond 7.5 MPa makes the AFF machine unstable due to the generation of excessive vibrations. Therefore, 7.5 MPa was identified as the limiting value of the extrusion pressure for further experiments.

Table 6.2: Result of the Phase-2 experiment of Stage-1

Exp. No.	Extrusion pressure '*P*' (MPa)	Spur gear		Straight Bevel gear	
		ΔR_a (μm)	ΔR_{max} (μm)	ΔR_a (μm)	ΔR_{max} (μm)
1	2.5	0.41	1.11	0.07	0.03
2	5	0.54	4.57	0.20	0.22
3	7.5	**0.76**	**8.32**	**0.54**	**5.34**
4	10	0.42	2.80	0.19	1.10

6.2 Findings from the Stage-2 Experiments

Twenty full-factorial experiments were conducted for the spur gear and straight bevel gear each in stage-2 to identify the optimum values of the finishing time and volumetric concentration of silicone oil, which is a measure of the viscosity of the AFF medium. The following subsections present the results of the stage-2 experimentation and their discussion.

6.2.1 Results and Discussion of Stage-2 Experiments

Table 6.3 presents the values of the considered responses for spur gear and straight bevel gear corresponding to each combination of variable parameters for the AFF process in the Stage-2 experimentation.

Table 6.3: Values of input parameters and the considered responses for spur and straight bevel gears in Stage-2 experiments (Petare and Jain, 2018a; 2018b)

Exp. No.	Input parameters		Response: Average (except radial runout) percentage reduction in (%)											
			Spur gear						*Straight bevel gear*					
	Finishing time 't' (minutes)	*Volumetric concentration of silicone oil 'O_c'*	*Total profile error 'PRF_a'*	*Total lead error 'PRF_β'*	*Total pitch error 'PRF_p'*	*Radial runout 'PRF_r'*	*Max. surface roughness 'PRR_{max}'*	*Avg. surface roughness 'PRR_a'*	*Single pitch error 'PRf_p'*	*Adjacent pitch error 'PRf_u'*	*Total pitch error 'PRF_p'*	*Radial runout 'PRF_r'*	*Max. surface roughness 'PRR_{max}'*	*Avg. surface roughness 'PRR_a'*
1	10	15	4.2	4.0	0.5	1.9	22.6	22.3	2.5	16.7	7.3	5.2	23.2	39.1
2	***25***	***10***	***21.4***	***40.8***	***15.9***	4.3	57.6	66.2	14.0	53.0	27.3	***67.7***	56.3	52.1
3	30	20	11.2	3.7	3.3	3.0	16.1	53.3	3.4	10.2	6.7	3.7	20.0	35.0
4	15	10	6.9	8.5	2.1	1.2	48.0	39.8	10.2	38.2	18.6	39.7	44.7	46.9
5	25	20	9.2	7.2	2.7	2.9	16.0	39.2	4.0	10.0	2.0	1.9	11.4	34.1
6	10	20	2.0	1.4	0.4	0.0	13.4	16.4	0.9	-8.2	-27.7	-6.9	8.2	26.1
7	20	25	6.2	1.3	0.6	0.7	3.3	9.7	0.4	-57.3	-43.4	-51.5	-14.1	15.9
8	20	20	6.3	2.8	2.2	1.2	15.3	33.7	3.2	2.9	-2.4	-2.3	9.8	31.0
9	15	20	5.1	2.4	0.6	0.3	13.7	20.6	1.3	2.8	-24.7	-3.5	8.2	30.9
10	30	25	10.9	2.1	0.2	2.1	12.5	11.1	0.9	-13.0	-35.3	-16.2	6.3	17.7
11	25	25	7.8	2.8	0.9	0.8	8.8	9.9	0.7	-27.8	-39.2	-17.4	-4.0	16.2
12	***30***	***10***	13.4	26.8	12.6	5.2	***57.9***	***67.4***	***21.5***	***58.6***	***29.6***	**47.8**	***62.1***	**58.4**
13	15	15	6.2	6.7	1.0	4.0	25.6	32.9	3.2	17.0	8.1	9.2	25.8	39.3
14	10	10	4.8	5.2	0.7	0.6	26.5	39.1	6.1	35.4	16.6	37.1	39.6	45.2
15	25	15	9.5	23.7	6.4	5.3	38.7	55.5	5.3	27.8	14.4	30.5	30.9	40.8
16	10	25	1.0	1.3	0.0	0.0	0.4	7.7	0.1	-90.8	-79.8	-65.5	-102.3	14.3
17	15	25	4.4	1.4	0.3	0.0	2.8	7.9	0.1	-67.1	-78.7	-55.0	-76.1	15.0
18	30	15	12.1	11.1	11.4	4.1	42.9	60.8	6.5	28.4	15.9	31.9	35.2	41.4
19	20	15	6.5	7.4	6.0	***6.1***	36.0	48.7	4.2	21.2	8.8	30.5	29.8	39.7
20	20	10	10.4	11.9	14.2	2.9	52.8	49.8	13.4	38.5	24.2	42	53.1	47.8

Figures 6.2 and 6.3 depict graphically the effects of the finishing time and volumetric concentration of silicone oil on the considered responses for spur and straight bevel gears, respectively, with the help of the regression equations obtained using the results of table 6.3 and the corresponding experimental values. The considered responses include: (i) avg. % reductions in total profile error *'PRF_a'* (fig. 6.2a) and in total lead error *'PRF_β'* (fig. 6.2b) for the spur gear only, (ii) avg % reductions in single pitch error *'PRf_p'* (fig. 6.3a) and in adjacent pitch error 'PRf_u' (fig. 6.3b) for the straight bevel gear only, (iii) avg. % reductions in total pitch error 'PRF_p' (fig. 6.2c for spur and 6.3c for straight bevel gear), in max. surface roughness 'PRR_{max}' (fig. 6.2e for spur and 6.3e for straight bevel gear), and in avg. surface roughness 'PRR_a' (fig. 6.2f for spur and 6.3f for straight bevel gear), and (iv) % reduction in radial runout 'PRF_r' (fig. 6.2d for spur and 6.3d for straight bevel gear).

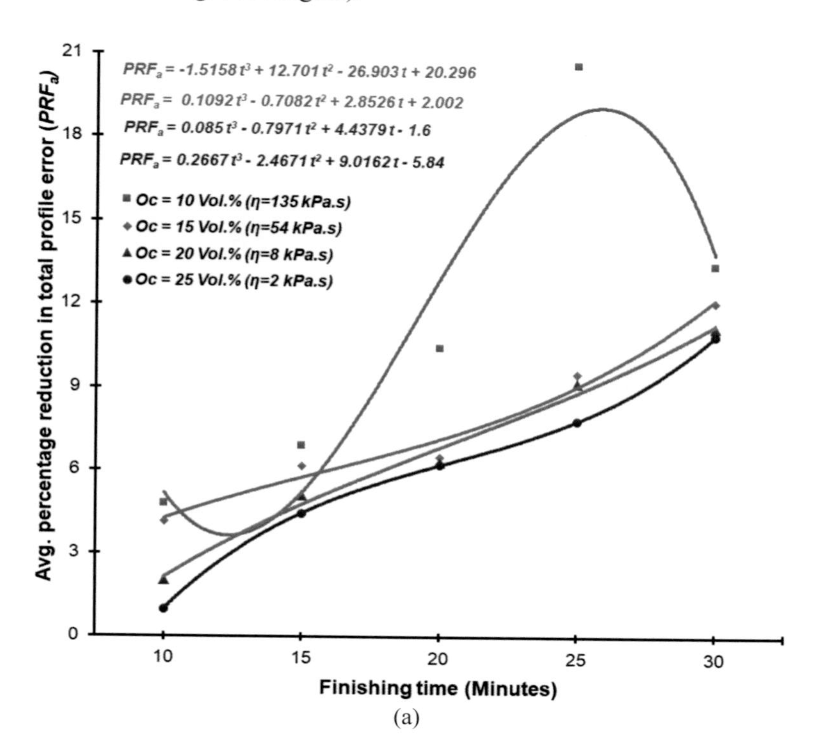

(a)

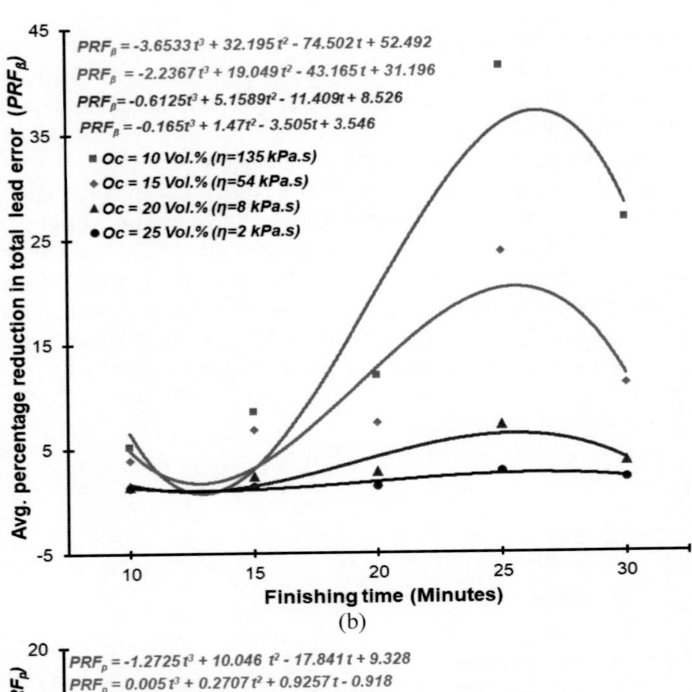
PRFβ = -3.6533t³ + 32.195t² - 74.502t + 52.492
PRFβ = -2.2367t³ + 19.049t² - 43.165t + 31.196
PRFβ= -0.6125t³ + 5.1589t² - 11.409t + 8.526
PRFβ = -0.165t³ + 1.47t² - 3.505t + 3.546
Oc = 10 Vol.% (η=135 kPa.s)
Oc = 15 Vol.% (η=54 kPa.s)
Oc = 20 Vol.% (η=8 kPa.s)
Oc = 25 Vol.% (η=2 kPa.s)
Avg. percentage reduction in total lead error (PRFβ)
45
35
25
15
5
-5
10
15
20
25
30
Finishing time (Minutes)

(b)

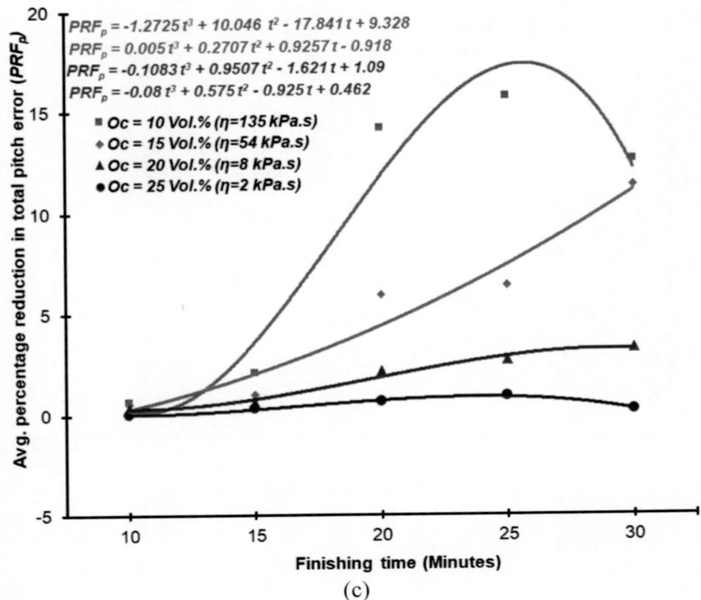
PRFp = -1.2725t³ + 10.046 t² - 17.841t + 9.328
PRFp = 0.005t³ + 0.2707t² + 0.9257t - 0.918
PRFp = -0.1083t³ + 0.9507t² - 1.621t + 1.09
PRFp = -0.08t³ + 0.575t² - 0.925t + 0.462
Oc = 10 Vol.% (η=135 kPa.s)
Oc = 15 Vol.% (η=54 kPa.s)
Oc = 20 Vol.% (η=8 kPa.s)
Oc = 25 Vol.% (η=2 kPa.s)
Avg. percentage reduction in total pitch error (PRFp)
20
15
10
5
0
-5
10
15
20
25
30
Finishing time (Minutes)

(c)

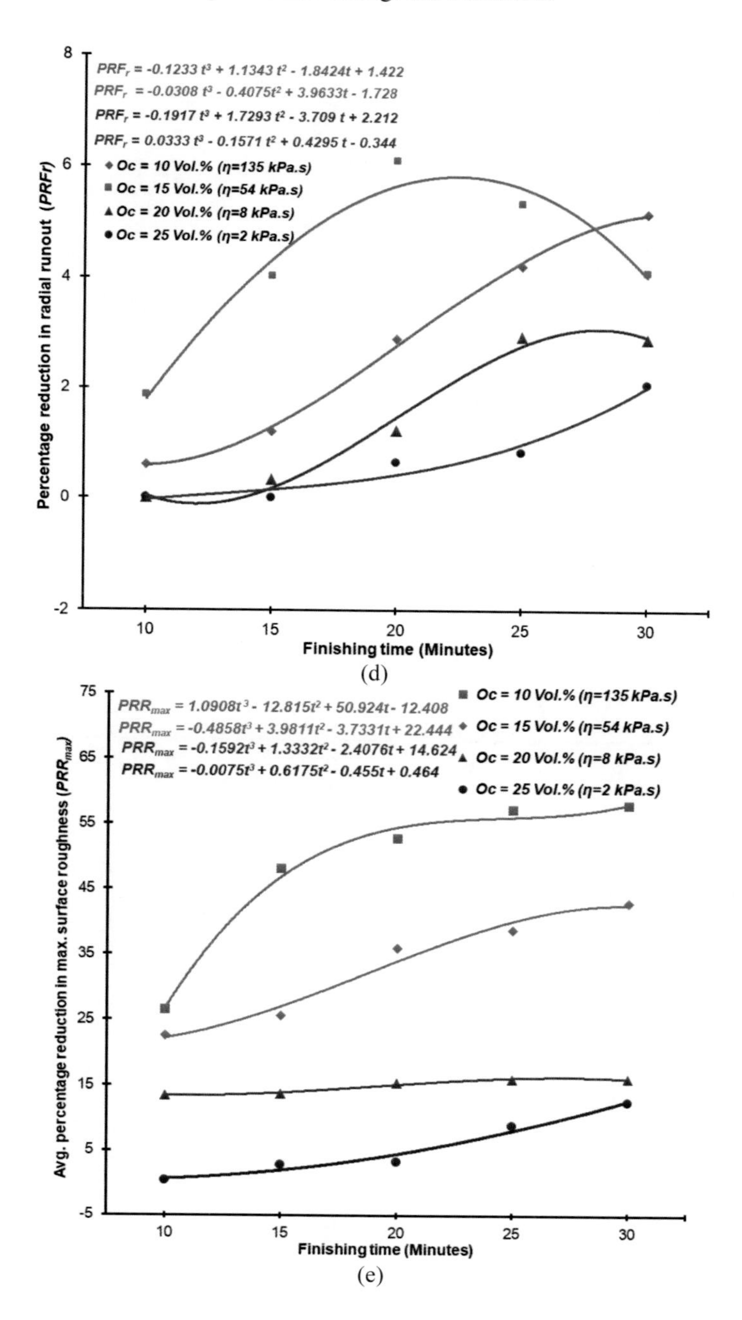
PRFr = -0.1233 t3 + 1.1343 t2 - 1.8424t + 1.422
PRFr = -0.0308 t3 - 0.4075t2 + 3.9633t - 1.728
PRFr = -0.1917 t3 + 1.7293 t2 - 3.709 t + 2.212
PRFr = 0.0333 t3 - 0.1571 t2 + 0.4295 t - 0.344
Oc = 10 Vol.% (η=135 kPa.s)
Oc = 15 Vol.% (η=54 kPa.s)
Oc = 20 Vol.% (η=8 kPa.s)
Oc = 25 Vol.% (η=2 kPa.s)
Percentage reduction in radial runout (PRFr)
Finishing time (Minutes)
PRRmax = 1.0908t3 - 12.815t2 + 50.924t - 12.408
PRRmax = -0.4858t3 + 3.9811t2 - 3.7331t + 22.444
PRRmax = -0.1592t3 + 1.3332t2 - 2.4076t + 14.624
PRRmax = -0.0075t3 + 0.6175t2 - 0.455t + 0.464
Oc = 10 Vol.% (η=135 kPa.s)
Oc = 15 Vol.% (η=54 kPa.s)
Oc = 20 Vol.% (η=8 kPa.s)
Oc = 25 Vol.% (η=2 kPa.s)
Avg. percentage reduction in max. surface roughness (PRRmax)
Finishing time (Minutes)

(d)

(e)

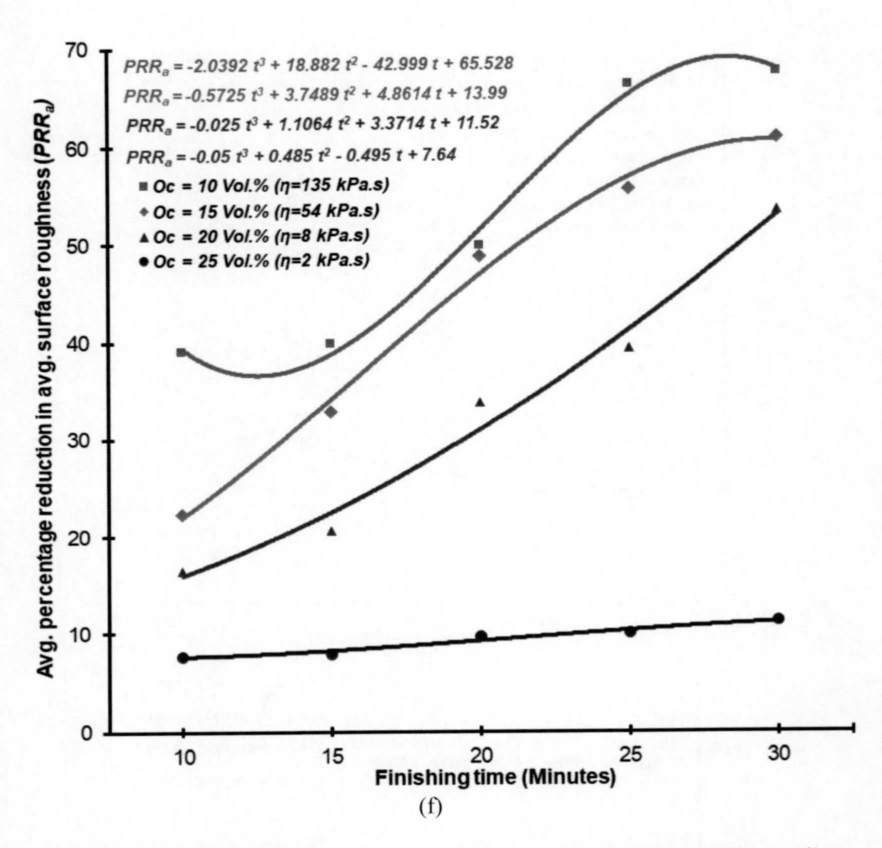

(f)

Fig. 6.2: Effect of the finishing time and volumetric concentration of silicone oil (i.e. viscosity of the AFF medium) on % reductions in *spur gear* (a) total profile error 'PRF_a', (b) total lead error 'PRF_β', (c) total pitch error 'PRF_p', (d) radial runout 'PRF_r', (e) maximum surface roughness 'PRR_{max}', and (f) average surface roughness 'PRR_a' (Petare and Jain, 2018a)

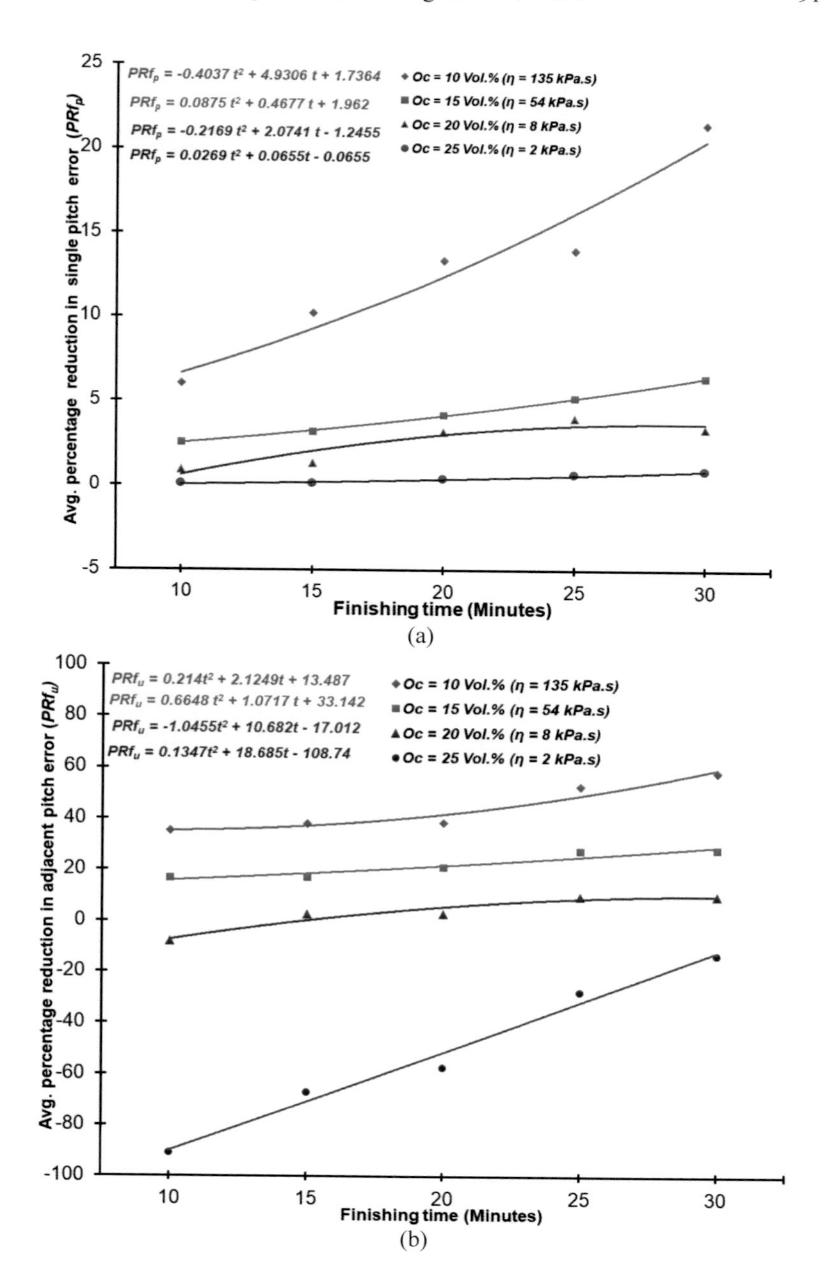

PRf_p = -0.4037 t² + 4.9306 t + 1.7364
PRf_p = 0.0875 t² + 0.4677 t + 1.962
PRf_p = -0.2169 t² + 2.0741 t - 1.2455
PRf_p = 0.0269 t² + 0.0655t - 0.0655
Oc = 10 Vol.% (η = 135 kPa.s)
Oc = 15 Vol.% (η = 54 kPa.s)
Oc = 20 Vol.% (η = 8 kPa.s)
Oc = 25 Vol.% (η = 2 kPa.s)
Avg. percentage reduction in single pitch error (PRf_p)
Finishing time (Minutes)
(a)
PRf_u = 0.214t² + 2.1249t + 13.487
PRf_u = 0.6648 t² + 1.0717 t + 33.142
PRf_u = -1.0455t² + 10.682t - 17.012
PRf_u = 0.1347t² + 18.685t - 108.74
Oc = 10 Vol.% (η = 135 kPa.s)
Oc = 15 Vol.% (η = 54 kPa.s)
Oc = 20 Vol.% (η = 8 kPa.s)
Oc = 25 Vol.% (η = 2 kPa.s)
Avg. percentage reduction in adjacent pitch error (PRf_u)
Finishing time (Minutes)
(b)

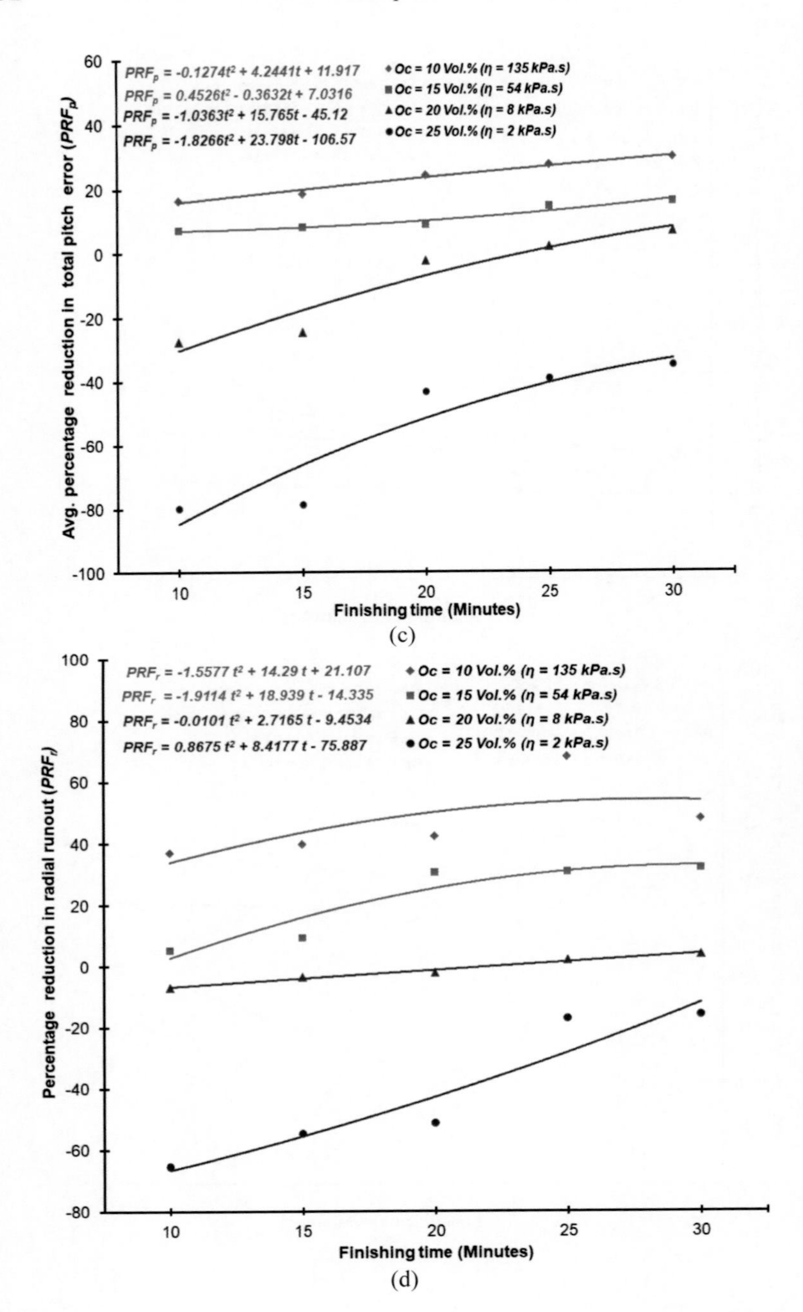
$PRF_p = -0.1274t^2 + 4.2441t + 11.917$
$PRF_p = 0.4526t^2 - 0.3632t + 7.0316$
$PRF_p = -1.0363t^2 + 15.765t - 45.12$
$PRF_p = -1.8266t^2 + 23.798t - 106.57$
Oc = 10 Vol.% (η = 135 kPa.s)
Oc = 15 Vol.% (η = 54 kPa.s)
Oc = 20 Vol.% (η = 8 kPa.s)
Oc = 25 Vol.% (η = 2 kPa.s)
Avg. percentage reduction in total pitch error (PRF_p)
60
40
20
0
-20
-40
-60
-80
-100
10
15
20
25
30
Finishing time (Minutes)

(c)

$PRF_r = -1.5577\ t^2 + 14.29\ t + 21.107$
$PRF_r = -1.9114\ t^2 + 18.939\ t - 14.335$
$PRF_r = -0.0101\ t^2 + 2.7165\ t - 9.4534$
$PRF_r = 0.8675\ t^2 + 8.4177\ t - 75.887$
Oc = 10 Vol.% (η = 135 kPa.s)
Oc = 15 Vol.% (η = 54 kPa.s)
Oc = 20 Vol.% (η = 8 kPa.s)
Oc = 25 Vol.% (η = 2 kPa.s)
Percentage reduction in radial runout (PRF_r)
100
80
60
40
20
0
-20
-40
-60
-80
10
15
20
25
30
Finishing time (Minutes)

(d)

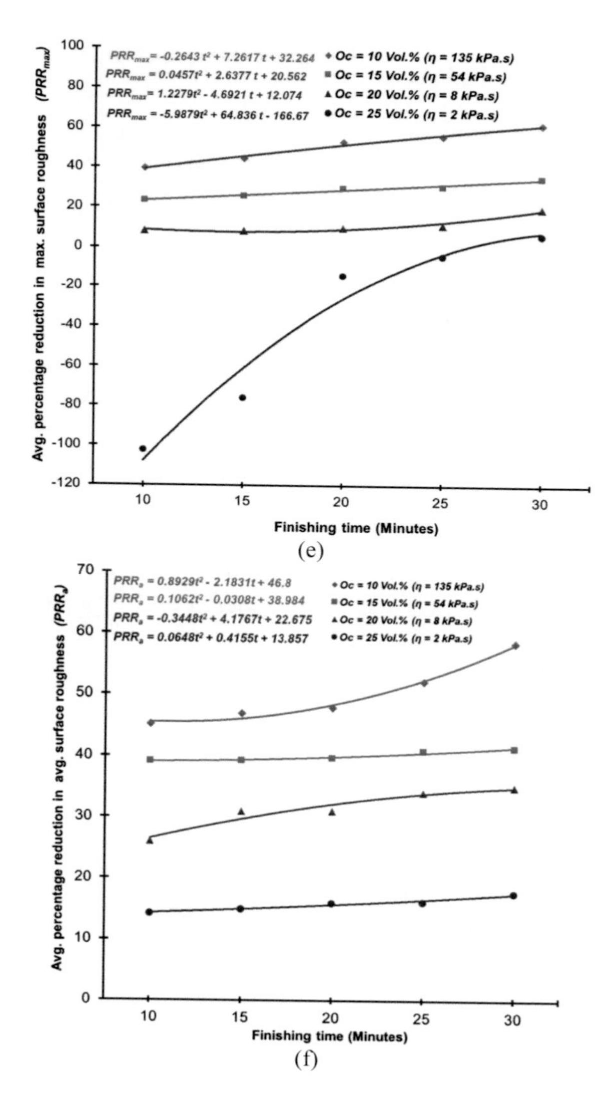

Fig. 6.3: Effect of the finishing time and volumetric concentration of silicone oil (i.e. viscosity of the AFF medium) on % reductions in *straight bevel gear* (a) single pitch error 'PRf_p', (b) adjacent pitch error 'PRf_u', (c) total pitch error 'PRF_p', (d) radial runout 'PRF_r', (e) maximum surface roughness 'PRR_{max}' and (f) average surface roughness 'PRR_a' (Petare and Jain, 2018b)

The following are the findings from table 6.3 and figs. 6.2 and 6.3, along with their explanations:

- The average percentage reductions in microgeometry errors and surface roughness parameters of the spur and straight bevel gears (i.e. PRF_{α}, PRF_{β}, PRf_{p}, PRf_{u}, PRF_{p}, PRF_{r}, PRR_{max}, and PRR_{a}) increase (i.e. microgeometry and surface finish improve) with an increase in the finishing time and with a decrease in volumetric concentration of silicone oil (i.e. increase in viscosity of the AFF medium) up to a certain value of finishing time. The increase in these parameters is smaller for the finishing time of 15-20 minutes than that for 20-30 minutes. This is due to the initial flattening of only higher surface peaks (caused by tooth cutter marks, burrs, etc.) on the flank surfaces of the unfinished spur and straight bevel gear teeth by the abrasive particles contained in the AFF medium without affecting the smaller surface peaks. As the finishing progresses, the abrasive particles start flattening other surface peaks also, which reduces peak-to-valley heights, giving higher values in percentage reductions in the considered responses. Consequently, their maximum values occur at 25 and 30 minutes of the finishing time for spur and straight bevel gear, respectively, and for a 10% vol. concentration of silicone oil (with the corresponding viscosity of the AFF medium being 135 kPa.s).
- Blending oil plays an important role in the proper mixing of the abrasive particles with putty and in holding them together against the extrusion pressure in the AFF process. Reducing its volumetric concentration below 10% to further increase the viscosity of the AFF medium can choke the cylinders containing the AFF medium and the restriction between the workpiece gear and its fixture. This results in insufficient slug formation due to poor bonding between the putty and abrasive particles.

It can be seen from table 6.3 that the combination of 25 minutes finishing time and 10% volumetric concentration of silicone oil (corresponding to experiment no. 2) for the AFF finished spur gear results in maximum values of average percentage reductions in its total profile error, total lead error, total pitch error, 2nd-highest value of avg. percentage reduction in max. and avg. surface roughness values, and 4th-highest value of percentage reduction in radial runout. On the other hand, the combination of a finishing time of 30 minutes and a volumetric concentration of silicone oil of 10% (corresponding to experiment no. 12) for the AFF-finished straight bevel gear yields maximum values of average percentage reductions in its single pitch error, adjacent pitch error, total pitch error, average surface roughness, maximum surface roughness, and the 2nd-highest percentage reduction in radial runout. Therefore, a 10% vol. concentration of silicone oil for both the spur and straight bevel gears and a finishing time of 25 minutes for the spur gears and 30 minutes for the straight bevel gears were identified as optimum values for further experiments. Gears finished with these combinations were considered the best finished spur and straight bevel gears in stage-2 experiments, and their evaluated surface roughness profiles, surface morphologies, wear characteristics, and microhardness are detailed in the following subsections.

6.2.2 Study of the Best Finished Gears in Stage-2 Experiments

6.2.2.1 Microgeometry Errors

Table 6.4 presents the values of the microgeometry errors of the unfinished and the best finished spur and straight bevel gears in the stage-2 experiment. Appendix C(i) presents their corresponding charts. It can be seen from these that for the stage-2 best finished *spur gear*, the AFF process reduced values of its (i) total profile error 'F_a' from 52.6 to 41.3 μm, improving its quality from DIN 11 to DIN 10; (ii) total lead error 'F_β' from 13.7 to 8.1 μm, enhancing its quality from DIN 9 to DIN 7; (iii) total pitch error 'F_p' from 106.6 to 89.6 μm, enhancing its quality from DIN 11 to DIN 10; and (iv) radial runout from 213.8 to 204.6 μm without any change in DIN quality. Similarly, for the stage-2 best-finished *straight bevel gear*, the AFF process reduced values of its (i) single pitch error 'f_p' from 82.9 to 65.4 μm without any change in DIN quality; (ii) adjacent pitch error 'f_u' from 110.2 to 45.6 μm, enhancing its quality from DIN 12 to DIN 9; (iii) total pitch error 'F_p' from 193.7 to 136.3 μm, enhancing its quality from DIN 11 to DIN 10; and (iv) radial runout 'F_r' from 130 to 67.9 μm, enhancing its quality from DIN 11 to DIN 9.

Table 6.4: Values of microgeometry errors of the unfinished and the best finished spur and straight bevel gears in the stage-2 experimentation (Petare and Jain, 2018a,b)

Microgeometry error (unit)	*Values for spur gear (DIN quality number)*		*Values for straight bevel gear (DIN quality number)*	
	Unfinished	*Stage-2 best finished*	*Unfinished*	*Stage-2 best finished*
Total Profile error 'F_a' (μm)	52.6 (11)	41.3 (10)	--	--
Total Lead error 'F_β'(μm)	13.7 (9)	8.1 (7)	--	--
Single Pitch error 'f_p'(μm)	--	--	82.9 (11)	65.4 (11)
Adjacent Pitch error 'f_u'(μm)	--	--	110.2 (12)	45.6 (9)
Total Pitch error 'F_p'(μm)	106.6 (11)	89.6 (10)	193.7 (11)	136.3 (10)
Radial Runout 'F_r' (μm)	213.8 (>12)	204.6 (>12)	130 (11)	67.9 (9)

6.2.2.2 Surface Roughness Profiles

Figures 6.4 and 6.5 present the surface roughness profiles of the unfinished and the best finished spur and straight bevel gear in the stage-2 experimentation. It can be noticed from fig. 6.4 that the AFF decreased the max. and avg. surface roughness values from 18.4 to 7.8 μm and from 1.7 to 0.5 μm, respectively, for the best finished *spur gear*. Similarly, the AFF decreased (fig. 6.5) max. and avg. surface roughness values from 14 to 5.3 μm and 1.7 to 0.7 μm, respectively, for best-finished *straight bevel gear*.

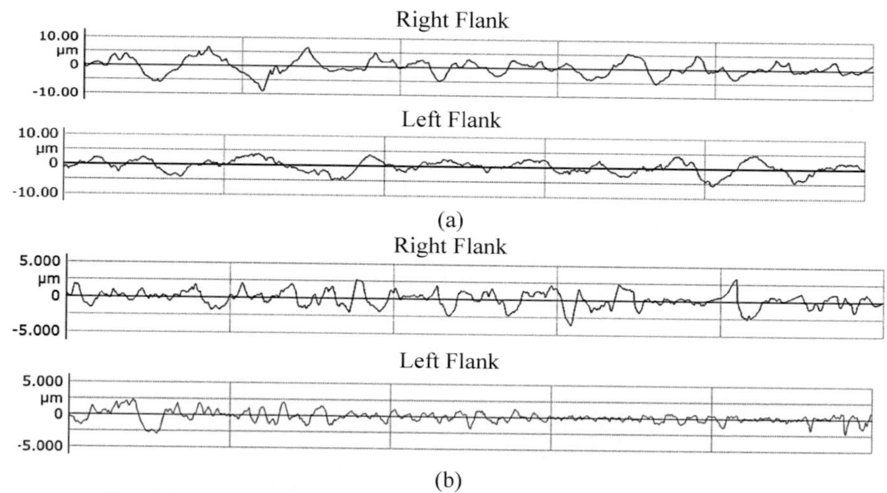

Fig. 6.4: Surface roughness profiles of the right and left flanks of the (a) unfinished spur gear, and (b) best finished spur gear in the Stage-2 experimentation (Petare and Jain, 2018a)

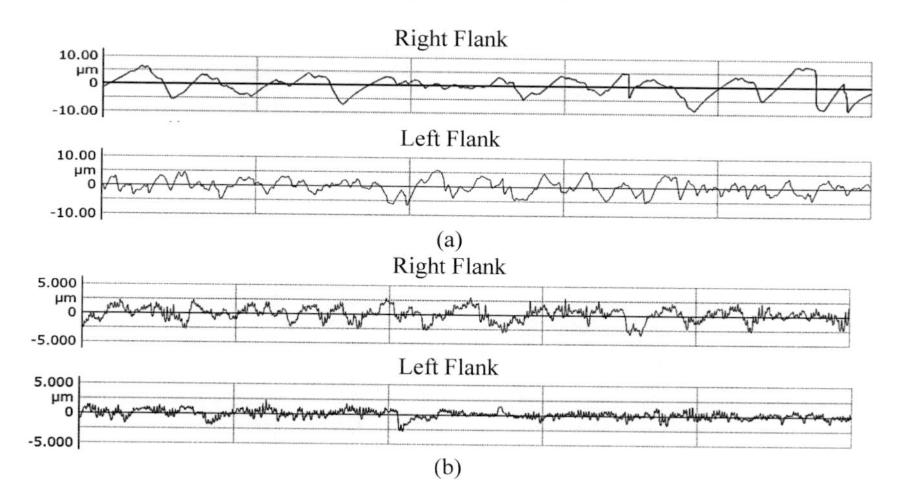

Fig. 6.5: Surface roughness profiles of the right and left flanks of the (a) unfinished straight bevel gear, and (b) best finished straight bevel gear in the Stage-2 experimentation (Petare and Jain, 2018b).

6.2.2.3 Surface Morphology

Figure 6.6 presents SEM images of the surface morphology of the flank surface of the unfinished and best-finished spur and straight bevel gears in the stage-2 experimentation. Marks of the gear cutter and surface roughness peaks are clearly visible in the surface morphology of the unfinished spur gear (fig. 6.6a) and straight bevel gear (fig. 6.6b), which are completely removed by the AFF process. It also gives a surface that is free of cutter marks, burrs, nicks, micro-cracks, and micro-pits, as shown in figs. 6.6c and 6.6d. The visible marks of the abrasive particles during the AFF process confirm that the material removal takes place due to abrading action followed by micro-cutting. It is also evident from the images that the AFF process is unable to completely remove the very deep cutter marks.

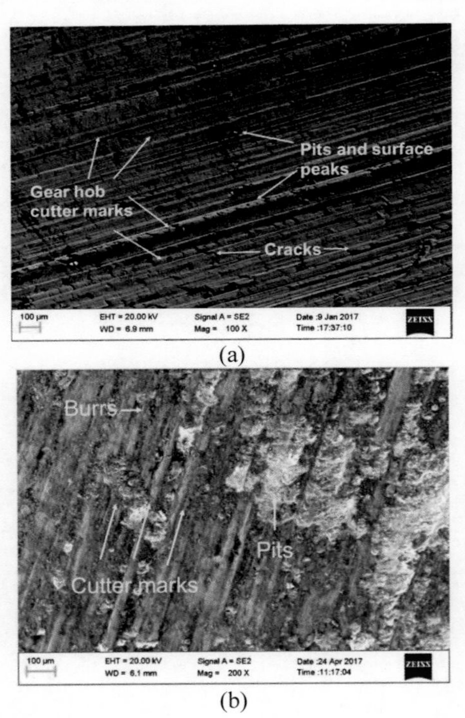

(a)

(b)

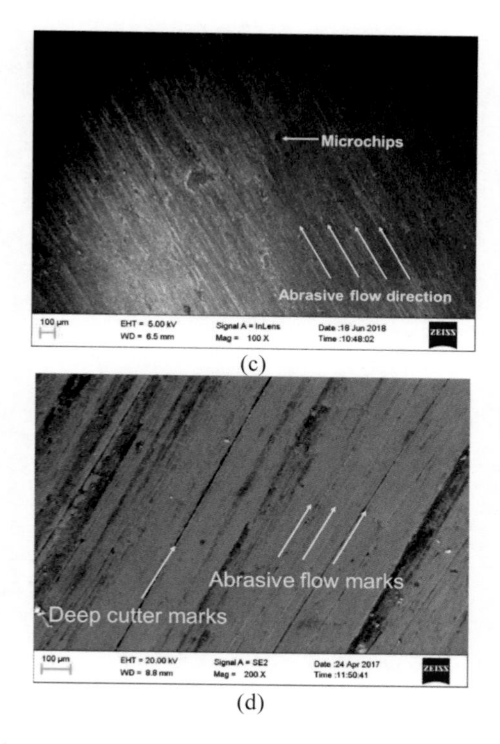

Fig. 6.6: SEM images showing the surface morphology of the flank surface of the (a) unfinished spur gear, (b) unfinished straight bevel gear, (c) stage-2 experimentation best-finished spur gear, and (d) stage-2 experimentation best-finished straight bevel gear (Petare and Jain, 2018a; and 2018b)

6.2.2.4 Wear Characteristics

Figures 6.7 and 6.8 depict the variation of the sliding friction force and the coefficient of sliding friction with time for the unfinished and the best-finished spur and straight bevel gear in the stage-2 experimentation. Table 6.5 presents the maximum values of the friction force and the coefficient of friction (obtained from the graphs of figs. 6.7 and 6.8), wear volume obtained from the fretting wear test, and computed values of specific wear rate and wear rate using Eq. 5.4.

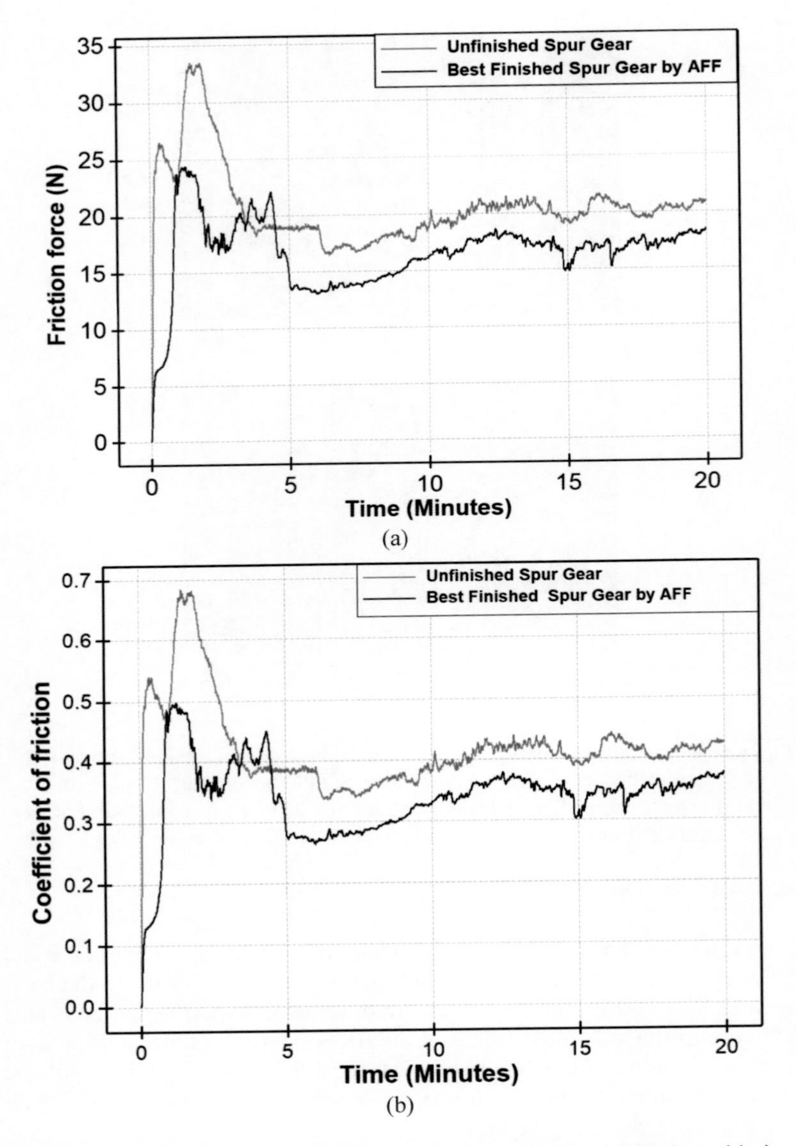

Fig. 6.7: Variation of the (a) friction force, and (b) coefficient of friction with time during the fretting wear test of the flank surface of the unfinished and the best finished spur gear in the stage-2 experimentation (Petare et *al.,* 2018)

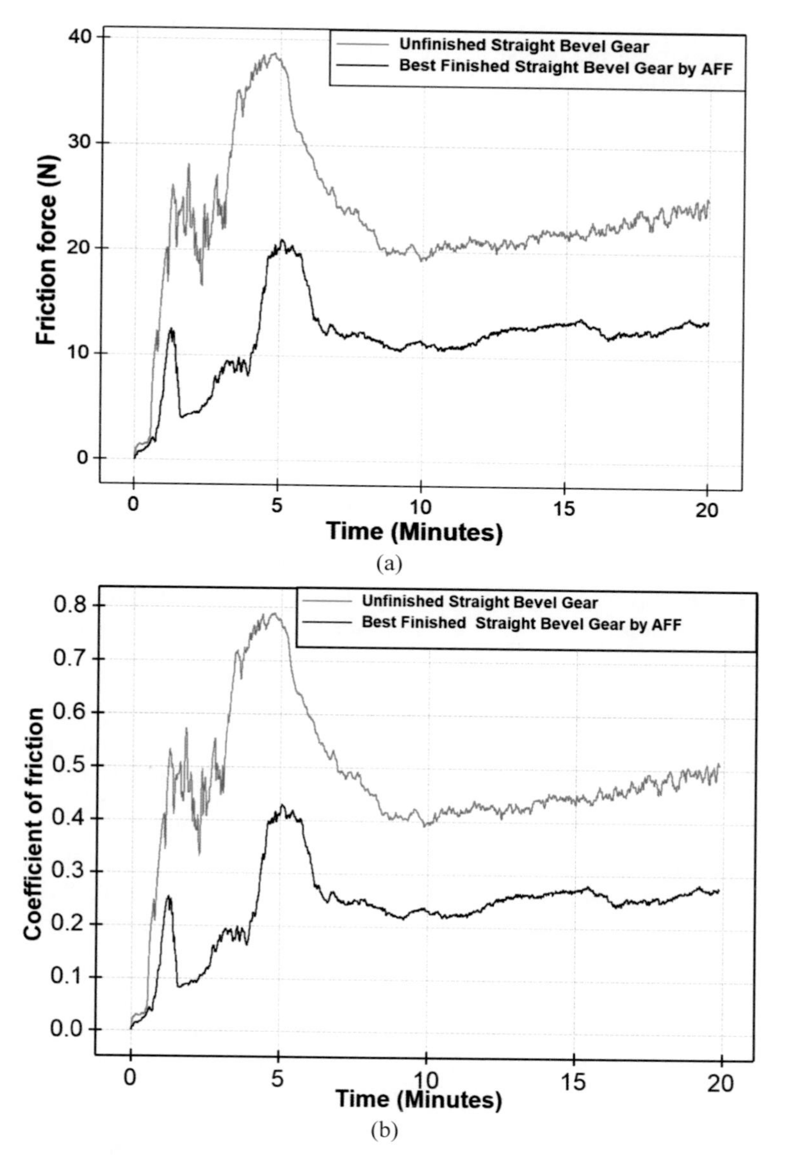

Fig. 6.8: Variation of the (a) friction force, and (b) coefficient of friction, with time during fretting wear test of tooth flank surface of the unfinished and the best finished straight bevel gear in the stage-2 experimentation

Table 6.5: Result of the fretting wear test for the unfinished and the best finished spur and straight bevel gears in the stage-2 experimentation (Petare and Jain, 2018b, Petare et *al.*, 2018)

Parameter name (unit)	*Unfinished spur gear*	*Stage-2 best finished spur gear*	*Unfinished straight bevel gear*	*Stage-2 best finished straight bevel gear*
Max. value of frictional force (N)	33.4	24.3	38.7	21.1
Max. value of coefficient of friction	0.7	0.5	0.8	0.4
Specific wear rate 'k' (mm^3/N-m)	14.8 x 10^{-6}	5.7 x 10^{-6}	15.5 x 10^{-6}	5.1 x 10^{-6}
Wear rate (mm^3/m)	7.1x 10^{-4}	2.9 x 10^{-4}	7.8 x 10^{-4}	2.6 x 10^{-4}
Wear volume 'V' (mm^3)	0.17	0.07	0.2	0.06

It can be observed from figs. 6.7 and 6.8 and table 6.5 that the finishing of spur and straight bevel gears by the AFF process significantly decreases the maximum values of the coefficient of friction and frictional force, wear volume, specific wear rate, and wear rate. Unfinished spur and straight bevel gears have large variations in their surface roughness profile, as shown in figs.6.4a and 6.5a. This means there is less contact area for the distribution of the applied force, resulting in a higher coefficient of friction and friction force and thus a higher wear rate or lower wear resistance. The best finished spur and straight bevel gears by the AFF process in the stage-2 experimentation have a smoother surface with very less surface roughness, which enhances the contact area for the distribution of the applied force and reduces frictional heat generation. This results in a lower coefficient of friction, friction force and wear volume, and enhanced wear resistance. This will increase their service life and mechanical efficiency (Pathak et al., 2015).

6.2.2.5 Microhardness

Figures 6.9 and 6.10 present the results of the microhardness evaluation for the unfinished and the best finished spur and straight bevel gears in the stage-2 experimentation. It can be observed from these figures that the microhardness of the best finished spur and straight bevel gears is more than their corresponding unfinished state due to the use of high extrusion pressure in the AFF process, which makes the AFF medium flow back and forth over the flank surfaces of the unfinished gears, thus removing material through abrasion. Axial and radial forces acting on the abrasive particles increase with the extrusion pressure. The continuous impact of the abrasive particles results in surface hardening, thereby increasing the microhardness of the AFF-finished components. It also develops compressive residual stress, which improves their fatigue strength (Sankar et al., 2009).

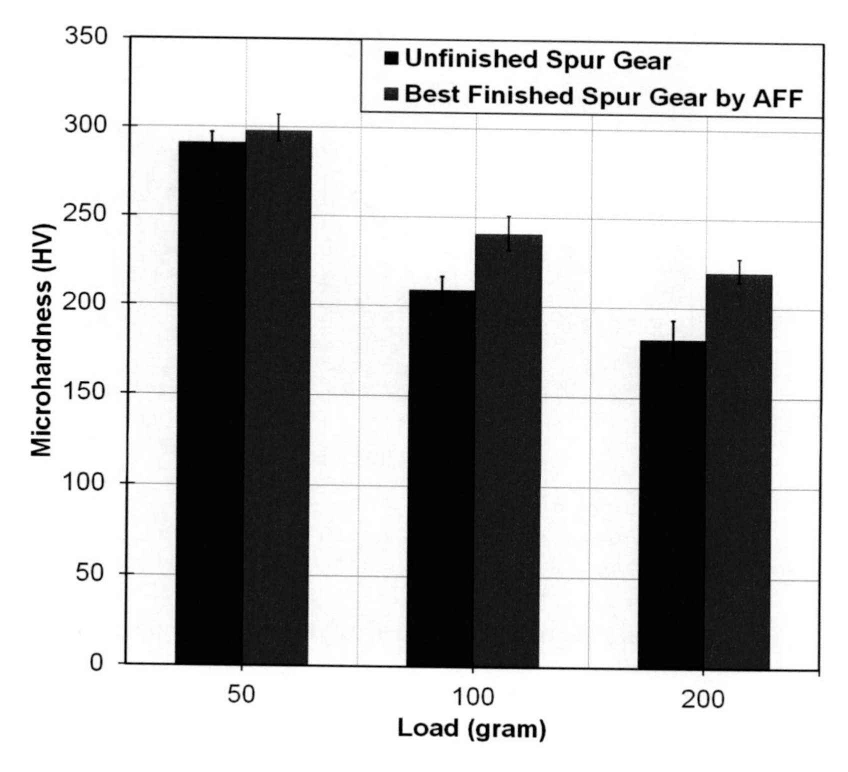

Fig. 6.9: Comparison of microhardness values for the unfinished and the best finished spur gear in the stage-2 experimentation

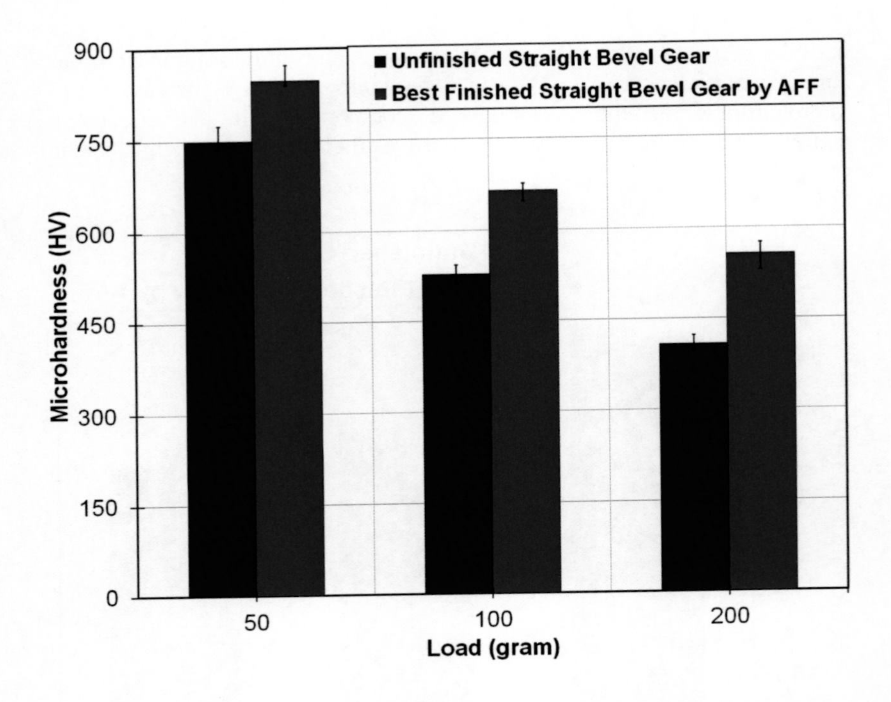

Fig. 6.10: Comparison of microhardness values for the unfinished and the best finished straight bevel gear in the stage-2 experimentation (Petare and Jain, 2018b)

6.3 Concluding Remarks from the Stage-2 Experiments

The following conclusions can be drawn from the results of stage-2 experiments:

- The finishing time and volumetric concentration of the blending oil (a measure of the viscosity of AFF medium) are found to have significant impact on the simultaneous improvement in surface finish, wear characteristics, and microgeometry of spur and straight bevel gears during their finishing by the AFF process.
- Reductions in surface roughness parameters and microgeometry errors continuously increase with the increase in viscosity of the AFF medium (i.e. decrease in vol.% of the blending oil) and finishing time, and attain their maximum values at a finishing time of 25 minutes (for spur gear) and 30 minutes (for straight bevel

gear) and 10% vol. concentration of the blending oil (i.e. 135 kPa.s as the viscosity of the AFF medium). They are identified as optimum values for further experiments.

- The surface morphology study of the AFF best finished spur and straight bevel gears revealed that their flank surfaces are free from cutter marks, burrs, nicks, micro-cracks, and micro-pits due to the restricted and concentrated movement of the AFF medium on the desired finishing areas. It also revealed that the finishing action by the AFF process is due to the abrading action followed by the micro-cutting on the gear teeth flank surfaces.
- The AFF process significantly reduced the maximum values of sliding friction force, coefficient of sliding friction, specific wear rate, wear rate, and sliding wear volume of spur and straight bevel gears.
- The AFF process increased the microhardness of spur and straight bevel gears significantly due to the use of a higher extrusion pressure, which exerts axial and radial forces to the abrasive particles, causing a work hardening of their flank surfaces and simultaneously imparting a very fine finish.

6.4 Findings from the Stage-3 Experiments

6.4.1 Results and Discussion of Stage-3 Experiments

Table 6.6 presents the values of the considered responses for spur and straight gears (i.e. avg. % reductions in total profile error 'PRF_a' and total lead error 'PRF_β' for *spur gear only*, and avg. % reductions in total pitch error 'PRF_p', max. surface roughness 'PRR_{max}', avg. surface roughness 'PRR_a', and % reduction in radial runout 'PRF_r' for both *spur and straight bevel gears*) for each combination of variable AFF parameters (extrusion pressure, volumetric concentration of abrasive particle, and size of abrasive particles) in the stage-3 experiments. Figures 6.11, 6.12, and 6.13 present graphs showing the variation in spur gear microgeometry errors (fig. 6.11) and surface roughness parameters (fig. 6.12), and straight bevel gear microgeometry errors and surface roughness parameters (fig. 6.13), with extrusion pressure (figs. 6.11a–6.13a), size of abrasive particles (figs. 6.11b–6.13b), and volumetric concentration of the abrasive particles (figs. 6.11c–6.13c) with the help of the regression equations obtained from the results of table 6.6 and the corresponding experimental values. The following inferences can be drawn from table 6.6 and figs. 6.11 to 6.13:

- It can be seen from table 6.6 that a combination of 7 MPa extrusion pressure, 120 mesh size of the abrasive particles, and 30% vol. concentration of the abrasive particles corresponding to exp. 10 for both the spur and straight bevel gears yielded the maximum values of their considered responses, which are highlighted through bold and italic text.
- Percentage reductions in microgeometry errors and surface roughness parameters of both spur and straight bevel gears continuously increase with the increase in extrusion pressure, and attain their maximum values at 7 MPa, as depicted in figs. 6.11a, 6.12a, and 6.13a. This is due to the increase in axial and radial forces acting on the abrasive particles contained in the AFF medium with the increase in extrusion pressure, which causes the shearing of more surface roughness peaks from the workpiece gear flank surfaces, thereby increasing reductions in its microgeometry errors and surface roughness parameters (Sankar et al., 2011).
- Percentage reductions in microgeometry errors and the surface roughness parameters of both spur and straight bevel gears continuously increase with the increase in the size of the abrasive particles, and attain their maximum value at 120 mesh (avg. diameter 127 μm), as illustrated in figs. 6.11b, 6.12b, and 6.13b. This is due to the less depth and width of the indentations caused by the higher mesh size of the abrasive particles (i.e. smaller abrasive particles), which gives lower surface roughness values and reduced microgeometry errors.
- Figures 6.11c, 6.12c, and 6.13c depict the existence of an optimum value of vol. concentration of the abrasive particles because percentage reductions in microgeometry errors and surface roughness parameters of both the spur and straight bevel gears increase up to 30% in value and then start decreasing with further increase in concentration. This can be explained by the fact that the number of abrasive particles available for the abrasion increases initially with an increase in their vol. concentration, which reduces surface roughness and microgeometry errors (i.e. increase % reductions in them) but increasing vol. concentration of abrasive particles beyond 30% in the AFF medium reduces the bonding between the putty and the abrasive particles. This causes a reduction in self-deformability of the AFF medium, thus lowering the reduction in microgeometry errors and surface roughness parameters (Jain and Adsul, 2000).

Table 6.6: Values of the input parameters and the considered responses for spur and straight bevel gears during stage-3 experiments

Exp. No.	*Input parameters*			*Responses: Average (except radial runout) percentage reduction in (%)*									
				Spur gear						*Straight Bevel gear*			
	Extrusion pressure 'P' (MPa)	*Size of abrasive particles 'M_a' (Mesh)*	*Vol. concentration of abrasive particles 'C_{av}'*	*Total profile error 'PRF_a'*	*Total lead error 'PRF_β'*	*Total pitch error 'PRF_p'*	*Radial runout 'PRF_r'*	*Avg. surface roughness 'PRR_a'*	*Max. surface roughness 'PRR_{max}'*	*Total pitch error 'PRF_p'*	*Radial runout 'PRF_r'*	*Avg. surface roughness 'PRR_a'*	*Max. surface roughness 'PRR_{max}'*
1	5	100	30	13.7	23.8	6.4	5.0	15.3	30.4	18.1	24.6	37.4	38.4
2	7	80	30	15.1	35.3	15.2	6.6	21.5	39.2	23.4	34.4	50.9	41.3
3	3	120	30	2.3	17.5	3.1	1.3	5.1	9.8	8.8	11.1	29.1	26.4
4	5	100	30	8.6	19.2	5.4	4.2	13.5	26.1	17.0	23.1	35.1	36.8
5	3	100	40	1.9	15.9	2.5	0.8	3.9	6.8	4.2	10.1	28.7	26.0
6	5	120	40	14.9	31.3	11.9	5.2	18.4	34.0	22.1	34.0	47.0	40.8
7	7	100	40	18.1	38.7	19.5	8.9	33.6	45.7	24.6	44.2	55.9	45.5
8	3	100	20	1.4	13.8	2.2	0.4	3.9	4.8	3.4	9.9	26.2	25.0
9	5	100	30	12.9	21.1	4.2	4.6	14.5	30.3	19.6	16.9	32.4	36.0
10	*7*	*120*	*30*	*22.7*	*42.0*	*20.8*	*10.0*	*38.5*	*57.3*	*28.7*	*51.5*	*62.1*	*50.4*
11	5	80	40	6.45	18.4	3.2	2.9	11.0	14.9	13.9	15.5	32.3	35.4
12	5	80	20	4.2	17.9	3.2	1.9	9.7	12.1	13.8	14.2	29.4	33.9
13	5	100	30	7.6	18.7	3.8	3.6	12.9	24.8	17.9	20.5	32.5	36.2
14	5	120	20	13.9	27.6	9.3	5.1	15.4	31.5	20.0	32.4	41.3	38.4
15	3	80	30	0.3	7.7	1.6	0.1	3.2	4.5	2.8	7.7	25.8	23.2
16	7	100	20	16.4	36.6	15.9	7.5	24.5	41.5	23.8	39.3	52.5	41.6
17	5	100	30	7.4	18.5	3.4	2.9	12.3	21.1	14.4	24.5	37.0	37.9

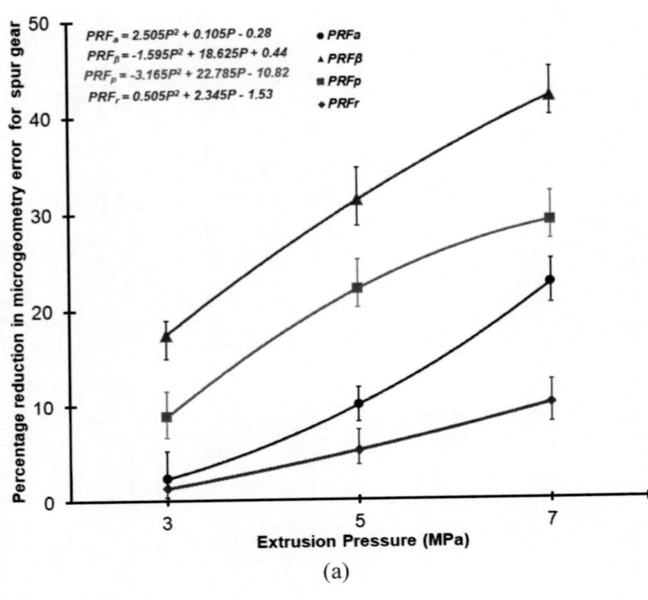
$PRF_a = 2.505P^2 + 0.105P - 0.28$
$PRF_\beta = -1.595P^2 + 18.625P + 0.44$
$PRF_p = -3.165P^2 + 22.785P - 10.82$
$PRF_r = 0.505P^2 + 2.345P - 1.53$
PRFa
PRFβ
PRFp
PRFr
Percentage reduction in microgeometry error for spur gear
Extrusion Pressure (MPa)

(a)

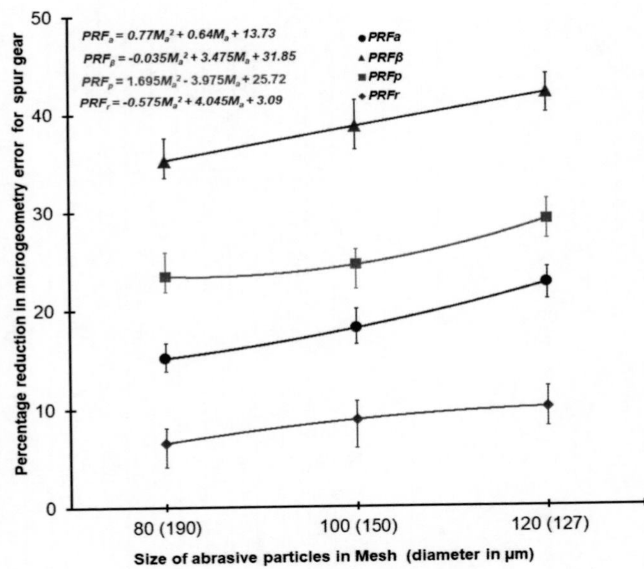
$PRF_a = 0.77M_a^2 + 0.64M_a + 13.73$
$PRF_\beta = -0.035M_a^2 + 3.475M_a + 31.85$
$PRF_p = 1.695M_a^2 - 3.975M_a + 25.72$
$PRF_r = -0.575M_a^2 + 4.045M_a + 3.09$
PRFa
PRFβ
PRFp
PRFr
Percentage reduction in microgeometry error for spur gear
80 (190)
100 (150)
120 (127)
Size of abrasive particles in Mesh (diameter in μm)

(b)

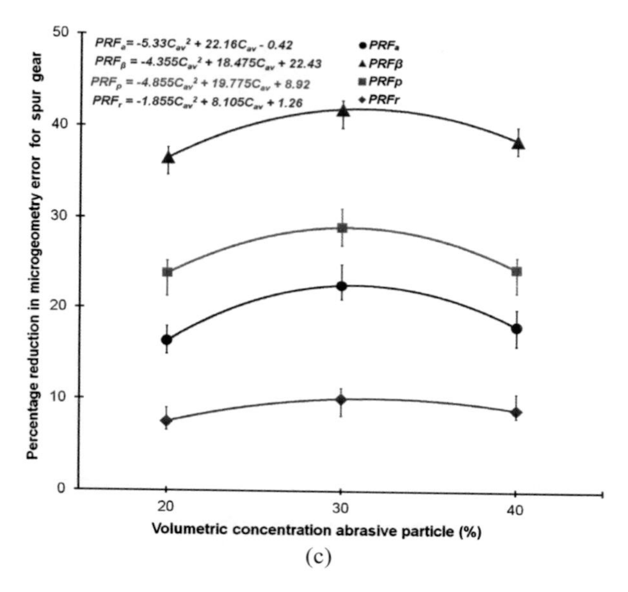

(c)

Fig. 6.11: Variation of percentage reduction in *microgeometry errors of spur gear* with (a) extrusion pressure, (b) size of abrasive particles, and (c) volumetric concentration of abrasive particles for stage-3 experiments (Petare et al., 2021)

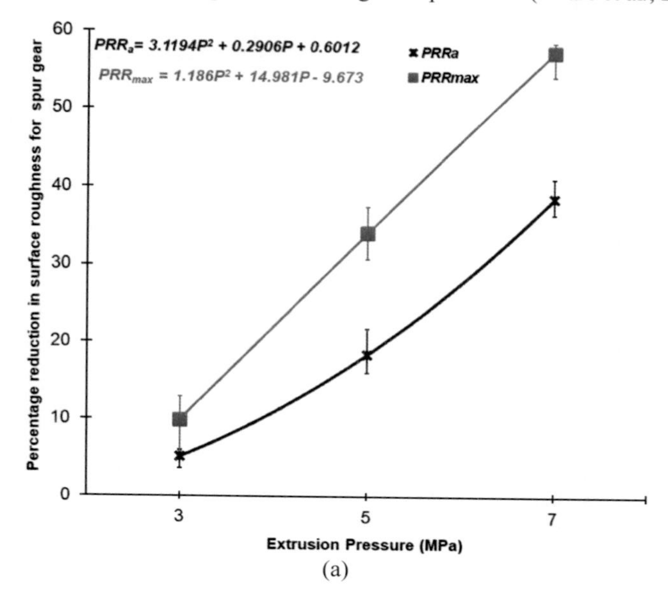

(a)

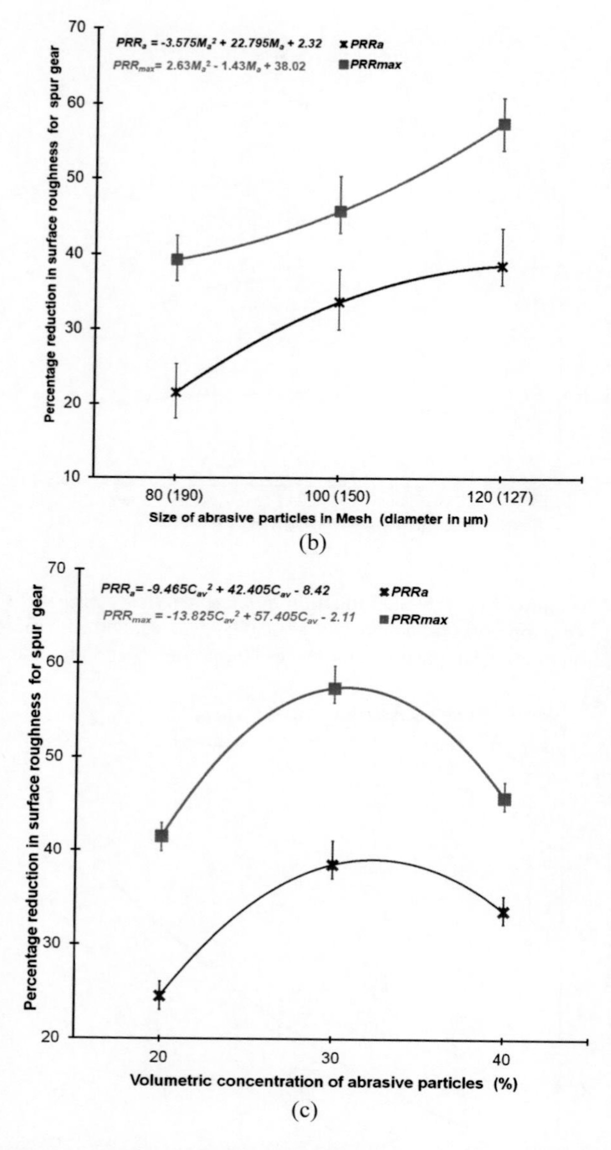

Fig. 6.12: Variation of percentage reduction in *surface roughness parameters* of *spur gear* with (a) extrusion pressure, (b) size of abrasive particles, and (c)

volumetric concentration abrasive particles for stage-3 experiments (Petare et al., 2021)

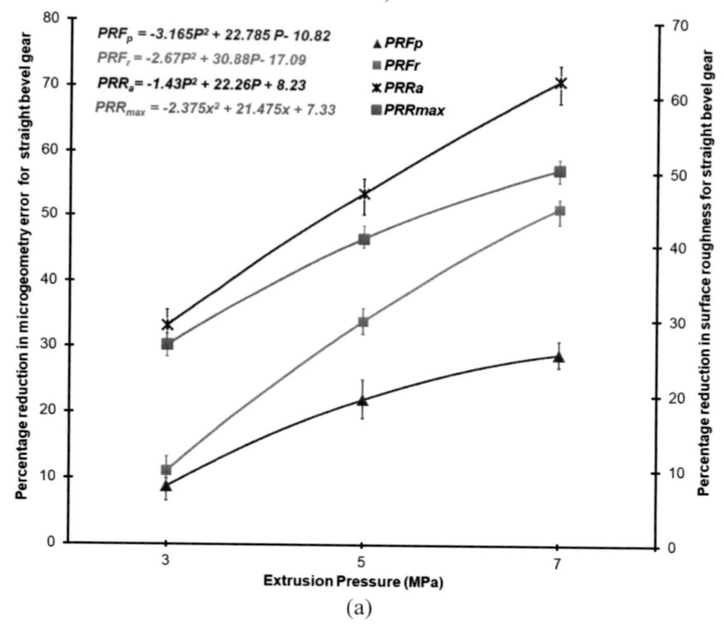

(a)

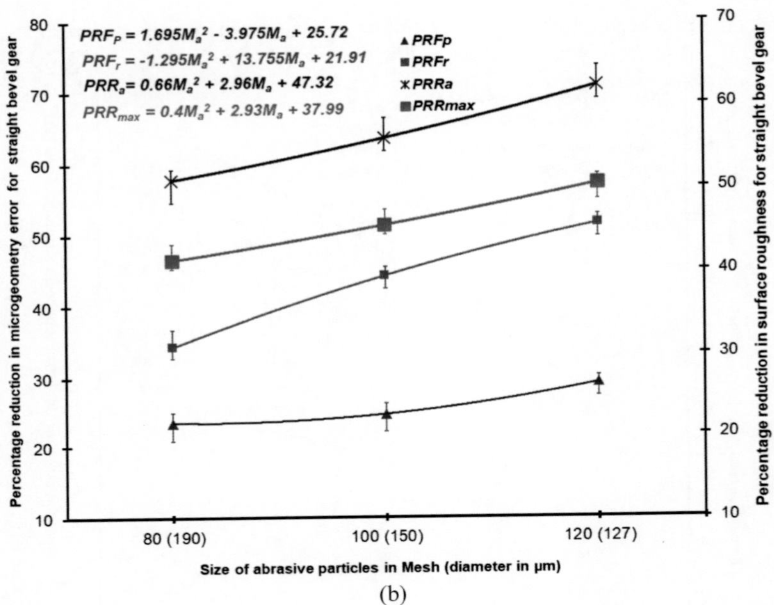

$PRF_P = 1.695M_a^2 - 3.975M_a + 25.72$
$PRF_r = -1.295M_a^2 + 13.755M_a + 21.91$
$PRR_a = 0.66M_a^2 + 2.96M_a + 47.32$
$PRR_{max} = 0.4M_a^2 + 2.93M_a + 37.99$
PRFp
PRFr
PRRa
PRRmax
Percentage reduction in microgeometry error for straight bevel gear
Percentage reduction in surface roughness for straight bevel gear
80
70
60
50
40
30
20
10
70
60
50
40
30
20
10
80 (190)
100 (150)
120 (127)
Size of abrasive particles in Mesh (diameter in μm)

(b)

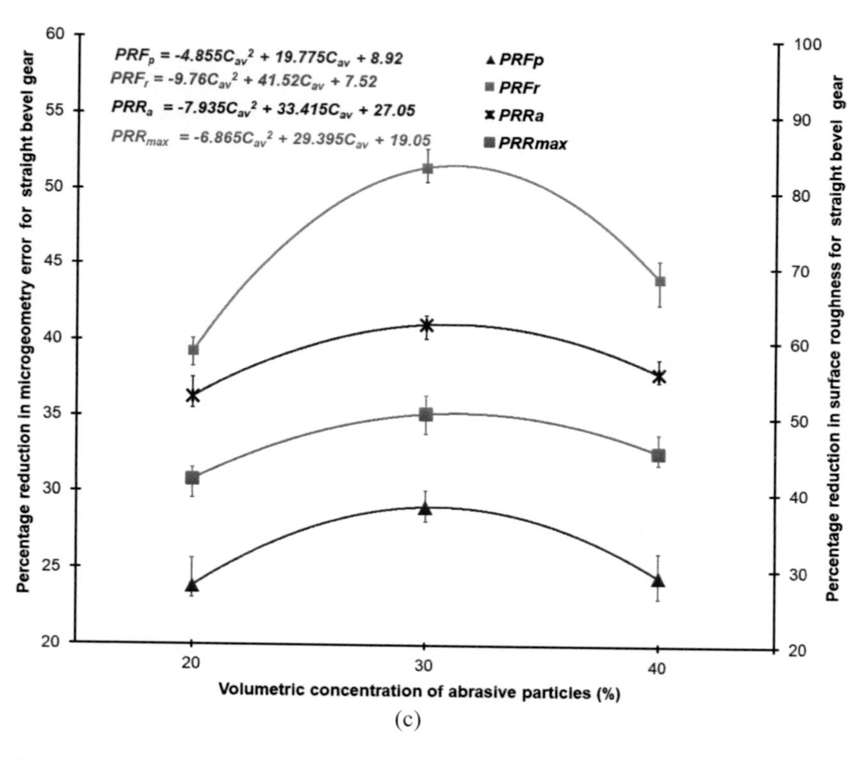

(c)

Fig. 6.13: Variation of percentage reduction in *microgeometry errors and surface roughness parameters* of *straight bevel gear* with (a) extrusion pressure, (b) size of abrasive particles, and (c) volumetric concentration abrasive particles for stage-3 experiments. (Petare et al., 2021)

6.4.2 Study of the Best Finished Gears in Stage-3 Experiments

The stage-3 experimentation identified 7 MPa as the extrusion pressure, 120 mesh as the size of the abrasive particles, and 30% volumetric concentration of the abrasive particles as the optimum values. The spur and straight bevel gears finished using this parametric combination are referred to as the best finished spur and straight bevel gears in the stage-3 experimentation, and whose surface roughness profiles, surface morphology, wear characteristics, and microhardness are studied and described in the following subsections.

6.4.2.1 Microgeometry Errors

Table 6.7 and Appendix C (ii) present the values of the microgeometry errors and the charts obtained from the CNC gear metrology machine for the unfinished and the best finished spur and straight bevel gears in the stage-3 experimentation. It can be observed that for the stage-3 best finished *spur gear*, the AFF process reduced its values of (i) total profile error 'F_a' from 56.9 to 44 µm, enhancing its quality from DIN 12 to DIN 11; (ii) total lead error 'F_β' from 7.2 to 4.2 µm, enhancing its quality from DIN 7 to DIN 5; (iii) total pitch error 'F_p' from 56.5 to 44.8 µm, enhancing its quality from DIN 10 to DIN 9; and (iv) radial runout 'F_r' from 54.7 to 49.2 µm, without any change in DIN quality. Similarly, for the stage-3 best finished *straight bevel gear*, the AFF process reduced values of its (i) total pitch error 'F_p' from 111 to 79.1 µm, enhancing its quality from DIN 10 to DIN 9; and (ii) radial runout 'F_r' from 108.7 to 52.7 µm, enhancing its quality from DIN 11 to DIN 9.

Table 6.7: Microgeometry errors for the unfinished and the best finished spur and straight bevel gears in stage-3 experimentation (Petare et al., 2021)

Microgeometry error (unit)	*Values for spur gear (DIN quality number)*		*Values for straight bevel gear (DIN quality number)*	
	Unfinished	*Stage-3 best finished*	*Unfinished*	*Stage-3 best finished*
Total Profile error 'F_a' (µm)	56.9 (12)	44.0 (11)	--	--
Total Lead error 'F_β'(µm)	7.2 (7)	4.2 (5)	--	--
Total Pitch error 'F_p'(µm)	56.5 (10)	44.8 (9)	111.0 (10)	79.1 (9)
Radial Runout 'F_r' (µm)	54.7 (10)	49.2 (10)	108.7 (11)	52.7 (9)

6.4.2.2 Surface Roughness Profiles

Figures 6.14 and 6.15 depict the surface roughness profiles of the unfinished and the best finished spur and straight bevel gear, respectively, in the stage-3 experimentation. It can be observed from fig. 6.14 that the AFF process reduced the max. and avg. surface roughness values from 16.1 to 6.8 μm and 1.2 to 0.7 μm, respectively, for the stage-3 best finished spur gear, and reduced the max. and avg. surface roughness values from 13 to 6.4 μm and 2.3 to 0.9 μm, respectively, for the stage-3 best finished straight bevel gear (fig. 6.15).

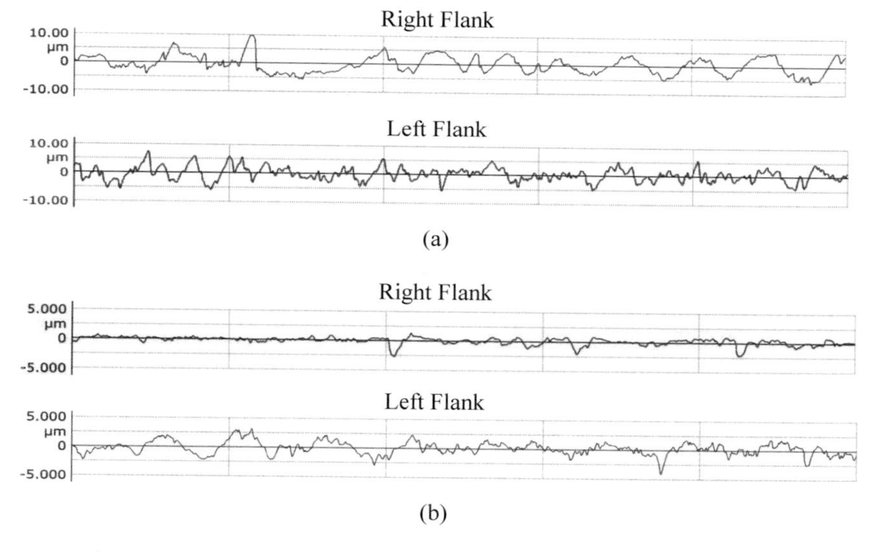

Fig. 6.14: Surface roughness profiles of the right and left flanks of the (a) unfinished spur gear, and (b) best finished spur gear in the stage-3 experimentation (Petare et al., 2021)

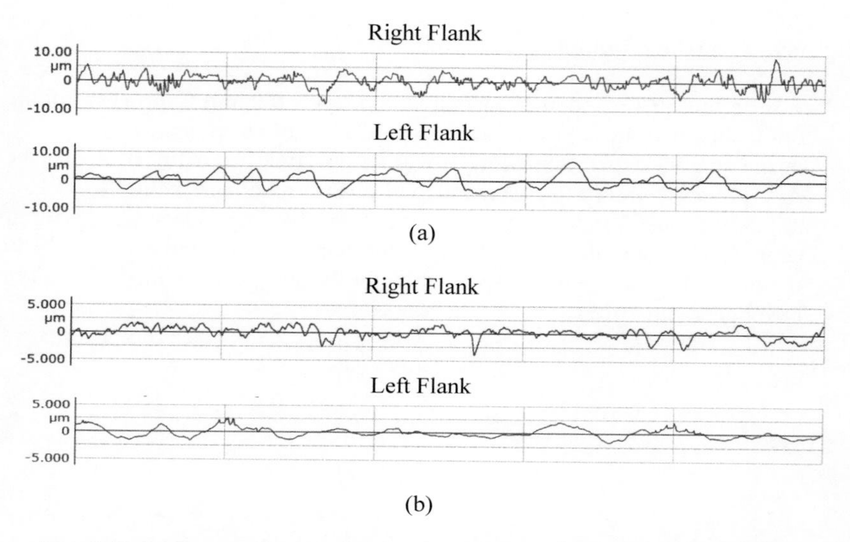

Fig. 6.15: Surface roughness profiles of the right and left flanks of the (a) unfinished straight bevel gear, and (b) best finished straight bevel gear in the stage-3 experimentation (Petare et al., 2021)

6.4.2.3 Surface Morphology

Figure 6.16 presents SEM images of the surface morphology of the flank surfaces of the unfinished and the best finished spur and straight bevel gears in the stage-3 experimentation. Marks made by the gear cutter, burrs, micro-chips, micro-pits, and surface roughness peaks can be seen on the flank surfaces of the unfinished spur gear (fig. 6.16a) and the unfinished straight bevel gear (fig. 6.16b). Their finishing by the AFF process completely removes the roughness peaks, micro-chips, and burrs, but it is unable to remove the deep marks of the gear cutter, pealed materials, and some micro-pits from the stage-3 best finished spur (fig. 6.16c) and straight bevel (fig. 6.16d) gears. The presence of pealed materials and marks of abrasive flow in these images indicates that micro-cutting with the plowing mode is the finishing mechanism of the AFF process.

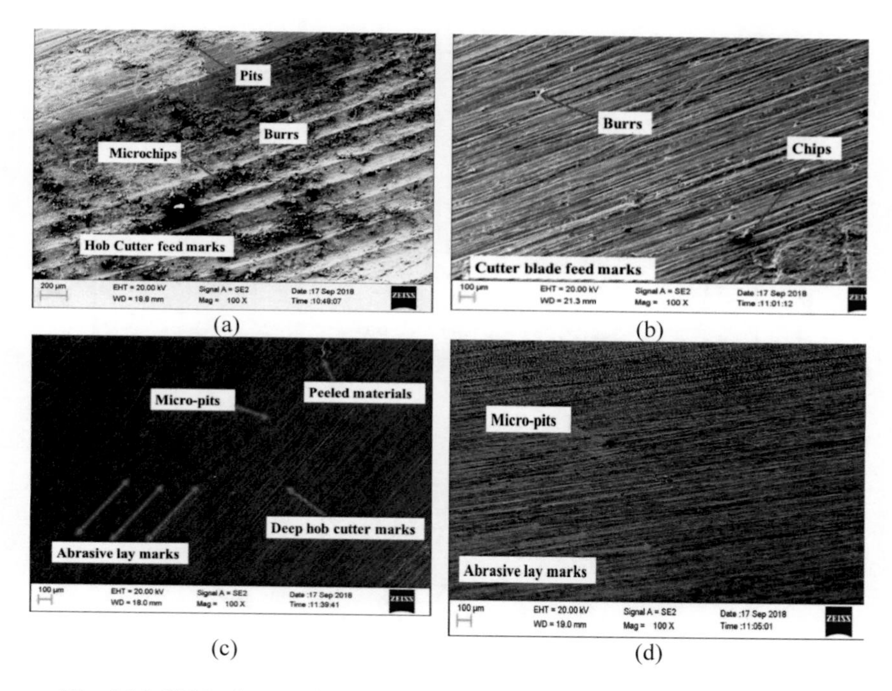

Fig. 6.16: SEM micrographs showing surface morphology of the flank surface of the (a) unfinished spur gear, (b) unfinished straight bevel gear, (c) best finished spur gear, and (d) best finished straight bevel gear, both in Stage-3 experimentation (Petare et al., 2021)

6.4.2.4 Wear Characteristics

Figures 6.17 and 6.18 depict the variation in the sliding friction force and the coefficient of sliding friction with time obtained from the tribometer for the flank surface of the unfinished and best finished spur gear and straight bevel gear, respectively, in the stage-3 experimentation. Table 6.8 presents a summary of the fretting wear test results, mentioning the maximum value of the friction force and the coefficient of friction (obtained from the graphs of figs. 6.17 and 6.18), the calculated values of specific wear rate, and the wear volume using Eq. 5.4 (refer to chapter 5).

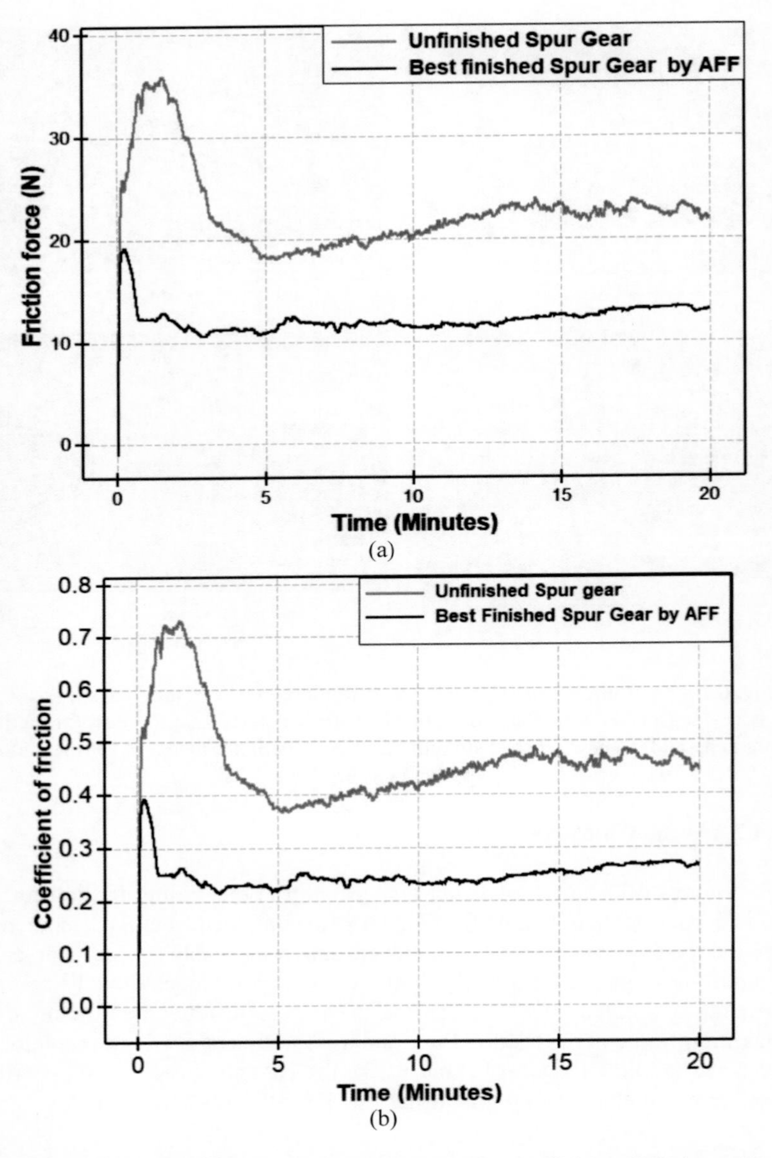

Fig. 6.17: Variation of the (a) friction force, and (b) coefficient of friction, with time during fretting wear test of the unfinished and best-finished spur gear in the Stage-3 experimentation (Petare et al., 2021)

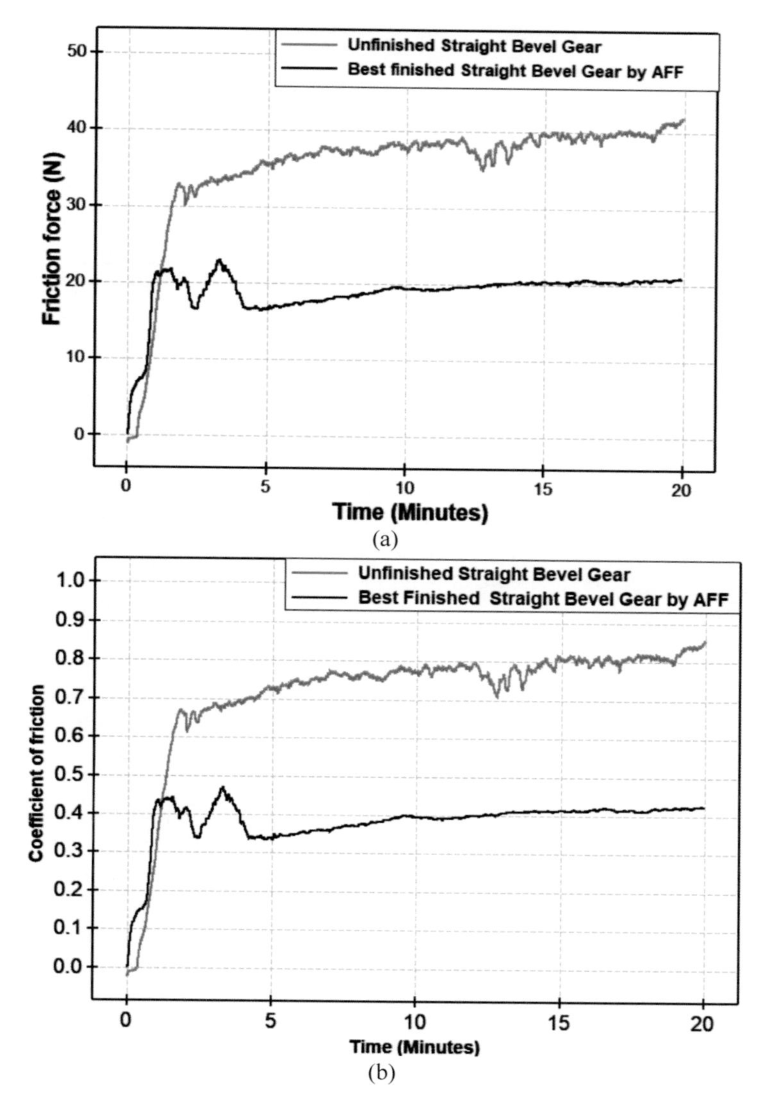

Fig. 6.18: Variation of the (a) friction force; and (b) coefficient of friction, with time during fretting wear test of the unfinished gear and best finished straight bevel gear in the Stage-3 experimentation (Petare et al., 2021)

Table 6.8: Results of fretting wear test for the unfinished and best finished spur and straight bevel gears in the Stage-3 experimentation (Petare et al., 2021)

Parameter name (unit)	*Spur gear*		*Straight Bevel gear*	
	Unfinished	*Stage-3 best finished*	*Unfinished*	*Stage-3 best finished*
Max. value of sliding friction force (N)	36.0	19.4	42	23.0
Max. value of coefficient of sliding friction	0.7	0.4	0.9	0.5
Specific wear rate 'k_i' (mm^3/N-m)	12.2 x 10^{-6}	5.3 x 10^{-6}	13.4 x 10^{-6}	6.3 x 10^{-6}
Wear rate (mm^3/m)	6.0 x 10^{-4}	2.6 x 10^{-4}	6.6 x 10^{-4}	3.1 x 10^{-4}
Wear volume 'V_i' (mm^3)	0.14	0.06	0.16	0.07

It can be observed from figs. 6.17 and 6.18 and table 6.8 that the friction force and the coefficient of sliding friction for flank surfaces of the unfinished spur and straight bevel gears increase sharply and attain maximum values in the initial 0 to 5 minutes of the wear test due to presence of higher surface roughness peaks. Thereafter they start decreasing due to the breakage of these higher surface roughness peaks and become stable once flatter surfaces are available. The initial increase in the friction force and the coefficient of sliding friction is much less in the case of the best finished spur and straight bevel gears due to the decreased heights of the surface roughness peaks, and thereafter it also becomes stable on the availability of smoother surfaces during the remaining wear test.

Figures 6.19 present the SEM images of the surface morphology after the fretting wear test for the flank surfaces of the unfinished spur gear (fig. 6.19a), unfinished straight bevel gear (fig. 6.19b), and best finished spur (fig. 6.19c) and best finished straight bevel gears in the stage-3 experimentation (fig. 6.19d). The SEM images (figs. 6.19a and 6.19b) depict the severely worn flank surfaces of the unfinished spur and straight bevel gear, revealing material displacement at different locations, surface cracks, sub-surface material displacement, pits, and accumulation of the worn debris. In contrast, the worn flank surfaces of the best finished spur and straight bevel gears in the stage-3 experimentation (figs. 6.19c and 6.19d) show much less worn debris and displacement of material and fewer pits. Material displacement and flaking in both gears indicate the scuffing wear of the gear flank surfaces.

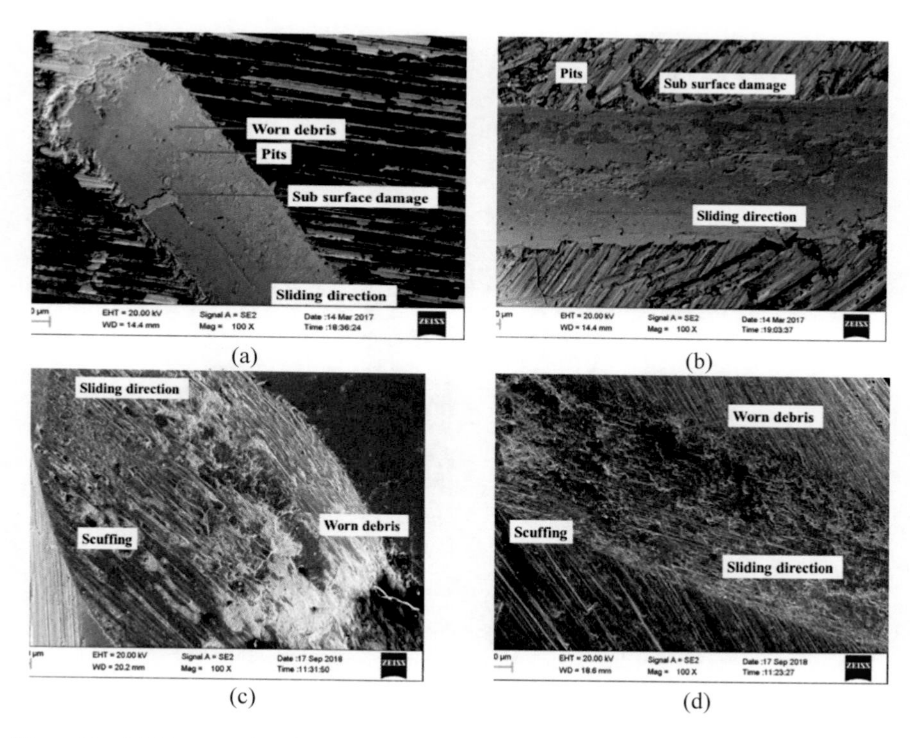

(a) (b) (c) (d)

Fig. 6.19: SEM images showing surface morphology after the fretting wear test for the flank surface of the (a) unfinished spur gear, (b) unfinished straight bevel gear, (c) best finished spur gear, and (d) best finished straight bevel gear in the stage-3 experimentation.

6.4.2.5 Microhardness

Figures 6.20 and 6.21 depict the results of the microhardness evaluation on flank surfaces of the (i) unfinished spur gear and best finished spur gear in the stage-3 experimentation (fig. 6.20); and (ii) unfinished straight bevel gear and best finished straight bevel gear in the stage-3 experimentation (fig. 6.21) using indentation forces of 50, 100, and 200 gm loads for a dwell time of 15s. Based on the graphs of figs. 6.20 and 6.21, it is clear that the microhardness of best finished spur and straight bevel gear is very high compared to their unfinished state for all the values of the indentation force. This is due to (i) an increase in extrusion pressure in the AFF process causing an exertion of more axial force on the abrasive particles thus more compressive force acting on them, which enhances work hardening and induces residual stresses; and (ii) finishing of the gear flank surfaces by the AFF process removes burrs, micro-cracks, and micro-pits and reduces surface roughness, therefore, more flat surface becomes available to resist indentation force compared to that in the unfinished gears, resulting in less indentation depth.

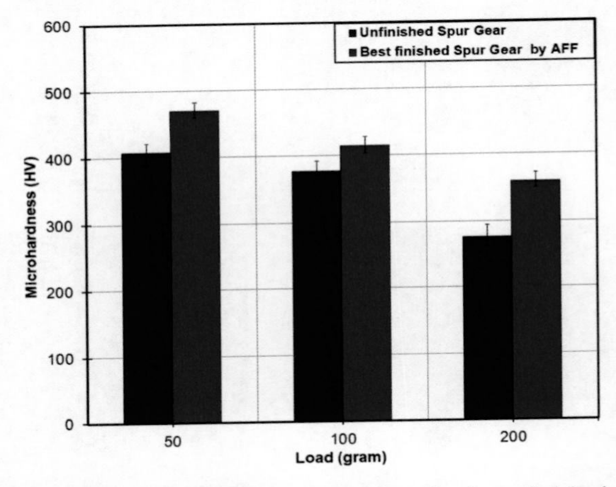

Fig. 6.20: Comparison of microhardness values for the unfinished and best finished spur gear in the stage-3 experimentation

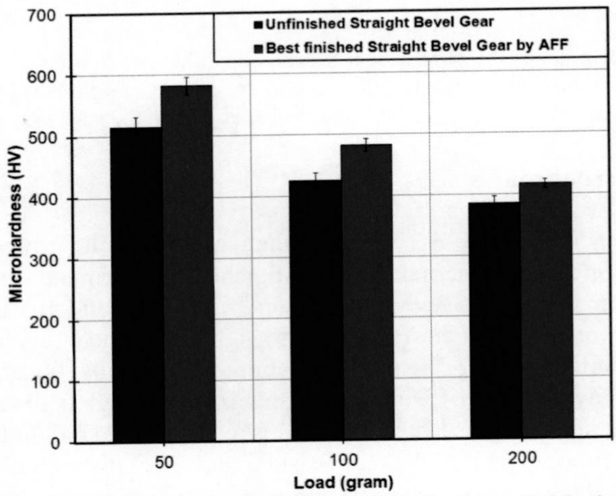

Fig. 6.21: Comparison of microhardness values for the unfinished and best finished straight bevel gear in the stage-3 experimentation

6.5 Concluding Remarks from the Stage-3 Experiments

The following conclusions can be drawn from the stage-3 experiments:

- Increases in the extrusion pressure and size of the abrasive particles continuously increase the reductions in microgeometry errors and surface roughness parameters of the spur and straight bevel gears and the attainment of their maximum values as 7 MPa and 120 Mesh, respectively.
- An optimum value of 30% exists for vol. concentration of the abrasive particles in the AFF medium because reductions in microgeometry errors and surface roughness parameters of the spur and straight bevel gears increase up to this value and then start decreasing. Therefore, 7 MPa extrusion pressure, 120 mesh size (127 μm avg. diameter) of the abrasive particles, and their 30% vol. concentration are identified as optimum parametric combination for the further study.
- Surface morphology of the spur and straight bevel gears shows that the AFF process completely removes roughness peaks, micro-chips, and burrs from their flank surfaces but is unable to remove some micro-pits, pealed materials, and deep marks of the gear cutter. The presence of the pealed materials and abrasive flow marks indicate that micro-cutting with the plowing mode is the finishing mechanism of the AFF process.
- Finishing of spur and straight bevel gears by the AFF process reduces the sliding friction force and the coefficient of sliding friction, which enhance their mechanical efficiency and service life. Surface morphology of flank surfaces of spur and straight bevel gears after their fretting wear test reveals material displacement and flaking, indicating their wear by the scuffing mode.
- The AFF process increases the microhardness of the finished spur and straight bevel gears due to higher extrusion pressure acting on the abrasive particles, which induces compressive residual stresses thus improving their fatigue life.

References

[1] Jain, V. K. and S. G. Adsul. 2000. "Experimental investigations into abrasive flow machining (AFM)." *International Journal of Machine Tools and Manufacture.* 40(7): 1003-1021. doi:10.1016/S0890-6955(99)00114-5

[2] Pathak, S, N. K. Jain, and I. A. Palani. 2019. *Finishing of Conical Gears by Pulsed Electrochemical Honing*. Newcastle upon Tyne, UK: Cambridge Scholars Publishing, ISBN: 1-5275-3366-2

[3] Petare, A. C. and N. K. Jain. 2018a. "Improving spur gear microgeometry and surface finish by AFF process." *Materials and Manufacturing Processes.* 33(9): 923-934. doi: 10.1080/10426914.2017.1376074

[4] Petare, A. C. and N. K. Jain. 2018b. "On simultaneous improvement of wear characteristics, surface finish and microgeometry of straight bevel gears by abrasive flow finishing process." *Wear.* 404-405: 38-49. doi: 10.1016/j.wear.2018.03.002

[5] Petare, A. C., A. Mishra, I. A. Palani and N. K. Jain. 2018. "Study of laser texturing assisted abrasive flow finishing for enhancing surface quality and microgeometry of spur gears." *The International Journal of Advanced Manufacturing Technology*.101(1):785-799. doi: 10.1007/s00170-018-2944-3

[6] Petare, A. C., N. K. Jain, and I. A. Palani. 2020. "Effect of finishing time on surface finish of spur gears by abrasive flow finishing (AFF) process." In *Advances in Unconventional Machining and Composites*. Singapore, Springer. 101-111. doi: 10.1007/978-981-32-9471-4

[7] Petare, A., N. K. Jain, and I. A. Palani. 2021. "On abrasive flow finishing of straight bevel gear." In *Lecture Notes in Intelligent Transportation and Infrastructure*. Singapore: Springer, 95-104. doi: 10.1007/978-981-33-4176-0_8

[8] Petare, A., N. K. Jain and I. A. Palani. 2021. "Simultaneous improvement of microgeometry and surface quality of spur and straight bevel gears by abrasive flow finishing process." *Journal of Micromanufacturing.* 4(2): 189-206. doi:10.1177/25165984211021010

[9] Sankar, M. R., V. K. Jain and J. Ramkumar. 2009. "Experimental investigations into rotating workpiece abrasive flow finishing." *Wear.* 267(1): 43-51. doi:10.1016/j.wear.2008.11.007

[10] Ravi Sankar, M., V. K. Jain, J. Ramkumar and Y. M. Joshi. 2011. "Rheological characterization of styrene-butadiene based medium and its finishing performance using rotational abrasive flow finishing process." *International Journal of Machine Tools and Manufacture.* 51(12): 947-957. doi:10.1016/j.ijmachtools.2011.08.012

CHAPTER SEVEN

PARAMETRIC OPTIMIZATION OF THE AFF PROCESS

The desirability function analysis (DFA) was used for the simultaneous optimization of the considered responses of the spur and straight bevel gears using the regression analysis models developed using the results of the stage-3 experiments presented in table 6.6

7.1 Formulation of the Optimization Models

Multi-response optimization using the DFA involves the conversion of each response into an individual desirability function 'd_{ij}' whose value ranges from 0 (when the response value is outside the considered range) to 1 (when the response is equal to its targeted value), i.e. $0 \leq d_{ij} \leq 1$; and then computation of the overall desirability function 'D_i' for the results corresponding to i^{th} experimental run using Eq. 7.1. It is the geometric mean of the individual desirabilities of all the responses.

$$D_i = \left(\prod_{j=1}^{j=n} d_{ij} \right)^{\frac{1}{\Sigma w_i}} \tag{7.1}$$

where, 'n' is total number of the considered responses; 'd_{ij}' is the desirability of the jth response for the experimental data of the ith run, and it is computed in the following manner according to type of response (Lee et al., 2018), which can be classified in the following three categories:

- **Smaller-the-better response or minimization of response**: It is used for the responses such as wear rate, wear volume, tool wear, surface roughness parameters, microgeometry errors, gear noise and vibrations, functional parameters of a gear, etc.

$$d_{ij} = \begin{cases} 1 & Y_{ij} < T_j \\ \left(\frac{U_j - Y_{ij}}{U_j - T_j}\right)^{w_j} & T_j \leq Y_{ij} \leq U_j \\ 0 & Y_{ij} > U_j \end{cases} \tag{7.2a}$$

- **Larger-the-better response or maximization of response**: It is used for the responses such as material removal rate (MRR), profit, strength of a weld joint or beam, tool life, percentage reduction in surface roughness parameters, etc.

$$d_{ij} = \begin{cases} 0 & Y_{ij} < L_j \\ \left(\frac{Y_{ij} - L_j}{T_j - L_j}\right)^{w_j} & L_j \leq Y_{ij} \leq T_j \\ 1 & Y_{ij} > T_j \end{cases} \tag{7.2b}$$

- **Target-is-the-best response:** It is used for the responses where a given target value is to be achieved, such as production rate, gear quality, load-carrying capacity of a gear, speed ratio between gear pair, etc.

$$d_{ij} = \begin{cases} 0 & Y_{ij} < L_j \\ \left(\frac{Y_{ij} - L_j}{T_j - L_j}\right)^{w_{j1}} & L_j \leq Y_{ij} \leq T_j \\ \left(\frac{U_j - Y_{ij}}{U_j - T_j}\right)^{w_{j2}} & T_j \leq Y_{ij} \leq U_j \\ 1 & Y_{ij} > U_j \end{cases} \tag{7.2c}$$

where, 'Y_{ij}' is the value of the jth response for the ith experimental data; 'T_j', 'L_j', and 'U_j' are the target values, acceptable lower limit, and acceptable upper limit of the jth response; and 'w_j', 'w_{j1}' and 'w_{j2}' are the weights assigned to the jth response. If 'w_j' is assigned a value equal to 1, then the desirability function becomes linear. Choosing a 'w_j' greater than 1 places more emphasis on the jth response, being close to the target value; and choosing $0 < w_j < 1$ makes a response less important. The combination of responses that gives the maximum overall desirability is considered the optimum parametric combination (Gupta and Jain, 2014).

In the present case, the concept of DFA was used to find the optimum values of three parameters of the AFF process (i.e. extrusion pressure, size of abrasive particles, and volumetric concentration of abrasive particles) through the simultaneous maximization of percentage reduction in the considered responses of microgeometry and surface roughness parameters of the spur and straight bevel gears (i.e. all the responses are the larger-the-better or the maximization type) and assigning equal weightage to each response. The following equations were used to compute the desirability value of each response of the spur and straight bevel gears considered during the stage-3 experiments for the ith combination of AFF parameters:

For spur gear $(DPRF_a^i) = \left[\frac{PRF_a - PRF_{a_{min}}}{PRF_{a_{max}} - PRF_{a_{min}}}\right]^{w_i}$ (7.3)

For spur gear $\left(DPRF_\beta^i\right) = \left[\frac{PRF_\beta - PRF_{\beta_{min}}}{PRF_{\beta_{max}} - PRF_{\beta_{min}}}\right]^{w_i}$ (7.4)

For spur and straight bevel gears $\left(DPRF_p^i\right) = \left[\frac{PRF_P - PRF_{P_{min}}}{PRF_{P_{max}} - PRF_{P_{min}}}\right]^{w_i}$ (7.5)

For spur and straight bevel gears $(DPRF_r^i) = \left[\frac{PRF_r - PRF_{r_{min}}}{PRF_{r_{max}} - PRF_{r_{min}}}\right]^{w_i}$ (7.6)

For spur and straight bevel gears $(DPRR_a^i) = \left[\frac{PRR_a - PRR_{a_{min}}}{PRR_{a_{max}} - PRR_{a_{min}}}\right]^{w_i}$ (7.7)

For spur and straight bevel gears $(DPRR_{max}^i) = \left[\frac{PRR_{max} - PRR_{max_{min}}}{PRR_{max_{max}} - PRR_{max_{min}}}\right]^{w_i}$ (7.8)

Where, $PRF_{a_{min}}$, $PRF_{\beta_{min}}$, $PRF_{P_{min}}$, $PRF_{r_{min}}$, $PRR_{a_{min}}$, $PRR_{max_{min}}$ and $PRF_{a_{max}}$, $PRF_{\beta_{min}}$, $PRf_{P_{max}}$, $PRF_{P_{max}}$, $PRF_{r_{max}}$, $PRR_{a_{max}}$,and $PRR_{max_{max}}$ are the minimum and maximum values of the percentage reduction in total profile error, total lead error, total pitch error, radial runout, maximum surface roughness, and average surface roughness respectively identified from the stage-3 experimental results from table 6.6. Therefore, for *spur gears*, the minimum value of PRF_a is set at 0.32%, with a maximum value 22.7%; for PRF_β the minimum value is set at 7.7%, with a maximum value 42%; for PRF_p the minimum value is set at 1.6%, with a maximum value 20.8%; for PRF_r the minimum value is set at 0.1%, with the maximum value 10%; for PRR_a the minimum value is set at 3.2%, with the maximum value 38.5%; and for PRR_{max} the minimum value is set at 4.5%, with the maximum value 57.3%. For *straight bevel gears,* PRF_p the minimum value is set at 2.8%, with the maximum value 28.7%; for PRF_r the minimum value is set at 7.7%, with the maximum value 51.5%; for PRR_a the minimum is set at 25.8%, with the maximum value 62.1%; and for PRR_{max} the minimum value is set at 23%, with the maximum value 50.4%. Since all the considered responses are to be maximized, they were considered as the *larger-the-better* and all of them were assigned equal weightage. This led to a weightage of 0.16 being assigned to each response of spur gear and 0.25 for straight bevel gear. The overall desirability for both spur and straight bevel gears were computed using the equations 7.9 and 7.10, respectively:

For Spur Gear:

$$D_i = \left[(DPRF_a^i)(DPRF_\beta^i)(DPRF_p^i)(DPRF_r^i)(DPRR_a^i)(DPRR_{max}^i)\right]^{0.16} \quad (7.9)$$

For straight bevel gear:

$$D_i = \left[(DPRF_p^i)(DPRF_r^i)(DPRR_a^i)(DPRR_{max}^i)\right]^{0.25} \quad (7.10)$$

Once the values of the overall desirability for all the experimental observations are calculated, then the parametric combination that gives the maximum possible value of overall desirability (i.e. either equal to 1 or nearest to 1) is identified as optimum. In the present case, the obtained values of the overall desirability for the spur and straight bevel gears are 1.0 and 0.99, respectively, and corresponding DFA-optimized values of extrusion pressure, size of the abrasive particles and volumetric concentration of the abrasive particles are: 6.7 MPa, 119.6 mesh, and 32.6% for the spur gears; and 6.9 MPa, 119.3 Mesh, and 34.4% for the straight bevel gears.

7.2 Experimental Validation of the Optimization Results

Three experiments were performed on spur and straight bevel gears each, and their average values were used to validate the results of optimization by the DFA and to compare the optimized values of the AFF process parameters and the considered responses with the optimum values identified by the RSM during stage-3 experiments. Stage-2 experimentation-identified optimum values of volumetric concentration of silicone oil (10%) and finishing time (25 and 30 minutes for the spur and straight bevel gear, respectively) were used in these experiments. Table 7.1 compares the results of validation experiments with those optimized by the DFA and the RSM-identified optimum values for the spur and straight bevel gears. Appendix C (iii) presents the microgeometry error values of the unfinished and finished spur and straight bevel gears using the optimized AFF process parameters. Figure 7.1 presents the bearing area or Abbot-Firestone curve for the unfinished (fig. 7.1a) and for the finished spur and straight bevel gears (fig. 7.1b) using the optimized AFF process parameters.

7.3 Concluding Remarks

The following conclusions can be drawn based on the result of the validation experiments:

- There is very good agreement between the optimum values of the considered responses obtained by the validation experiment with those predicted by the DFA and identified by the RSM (i.e. 7 MPa, 120 mesh, and 30% for both the spur and straight bevel gears). Prediction error for the DFA lies in the range between -9.7 and 10.3 for the spur gears, and between 1.7 and 8.9 for the straight bevel gear, i.e. the prediction error is higher for the spur gears than the straight bevel gears.
- The DFA overpredicts the values of % reductions in total profile error, total lead error, and total pitch error for the AFF-finished spur gear, and % reductions in total pitch error, radial runout, and avg. surface roughness for the AFF-finished straight bevel gear compared to those predicted by the validation experiments and RSM. RSM overpredicts the values of % reductions in the radial runout of the AFF-finished spur gear and the max. surface roughness of the AFF-finished straight bevel gear.
- The DFA underpredicts the % (i.e. negative prediction error) reductions in the max. and avg. surface roughness of AFF-finished spur gear, i.e. their values by the validation experiments are higher than those predicted values by the DFA. The DFA overpredicts (i.e. positive prediction error) the % reduction in microgeometry errors in the spur gear and all the considered responses for the straight bevel gear.
- It can be seen from fig. 7.1 that finishing by the AFF process using the optimized parameters makes the bearing area of the spur and straight bevel gears more uniform compared to their unfinished state. This is due to the reduction in the max. and avg. surface roughness of the spur gear from 15.8 to 6.3 μm and 1.6 to 0.8 μm, respectively, and from 13.6 to 7.4 μm and 2 to 0.7 μm for the straight bevel gear, respectively, by the AFF process. These reductions in surface roughness and consequent improvement in the bearing area of the spur and straight bevel gears will improve their wear characteristics, which will result in a better tribological performance during their service life.
- Validation experiments of the optimization result improved the spur gear quality in DIN standard from >12 to 11 for total profile

error, from 9 to 7 for total lead error, from 10 to 9 for total pitch error, and from 8 to 7 for radial runout. Similarly, the quality of the straight bevel gear improved from 9 to 8 for total pitch error and from 9 to 7 for radial runout.

- A controlled and predefined movement of abrasive particles in the AFF process removes cutter marks, burrs, pits, micro cracks, deep grooves, chips, and surface roughness peaks from the flank surfaces of spur and straight bevel gears, which improves their surface morphology after finishing by the AFF process.

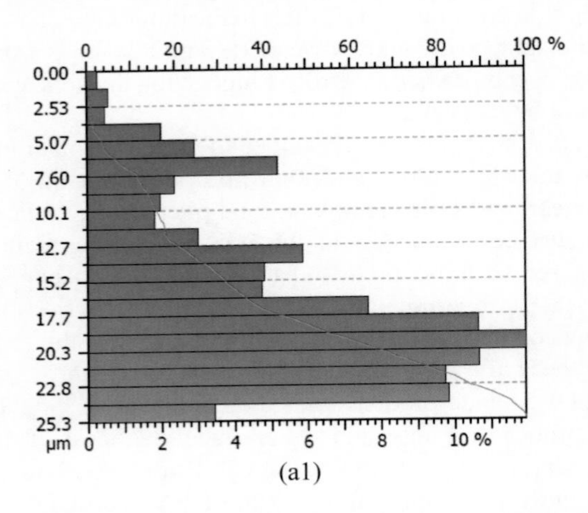

(a1)

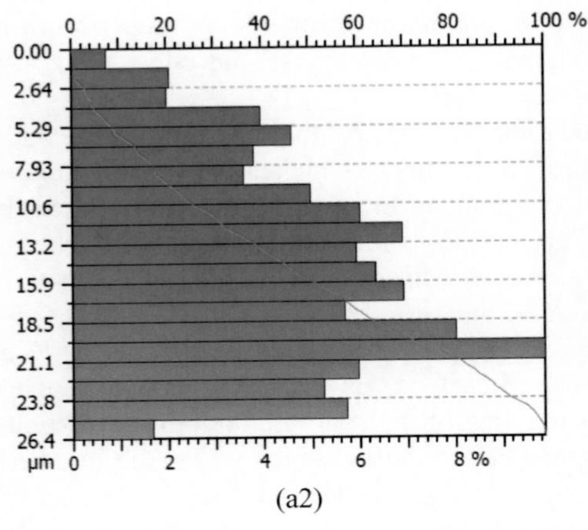

(a2)

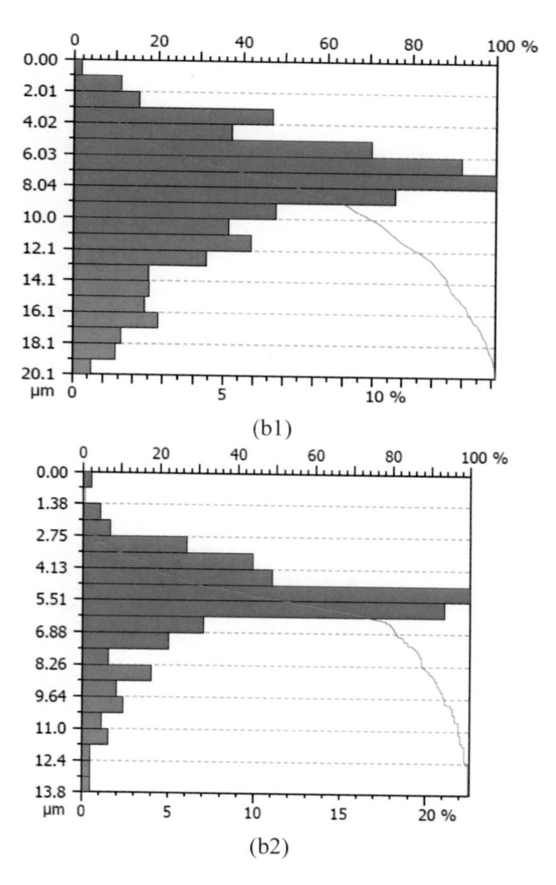

Fig. 7.1: Bearing area curve for the (a1) unfinished spur gear, (a2) unfinished straight bevel gear, (b1) AFF finished spur gear, and (b2) AFF finished straight bevel gear, using the optimum AFF process parameters given by the DFA (Petare et al., 2021).

Table 7.1: Comparison of the results of the validation experiment with the results of optimization by the DFA and the RSM-identified optimum results from stage-3 experiments

Approach	*Identified optimum values of process parameters*			*Optimum responses*																	
	Extrusion pressure (MPa)	*Size of abrasive particles (Mesh)*	*Vol. concentration of abrasive particles (%)*	*Total profile error (µm)*		*PRF_a (%)*	*Total lead error (µm)*		*PRF_β (%)*	*Total pitch error (µm)*		*PRF_p (%)*	*Radial runout (µm)*		*PRF_r (%)*	*Max. surface roughness (µm)*		*PRR_{max} (%)*	*Avg. surface roughness (µm)*		*PRR_a (%)*
				Before AFF	*After AFF*		*Before AFF*	*After AFF*		*Before AFF*	*After AFF*		*Before AFF*	*After AFF*		*Before AFF*	*After AFF*		*Before AFF*	*After AFF*	
For spur gear																					
RSM (Stage-3)	7	120	30	56.9	44	*22.7*	7.2	4.2	*41.7*	56.5	44.8	*20.7*	54.7	49.2	*10.1*	16.1	6.8	*57.4*	1.2	0.8	*38.6*
DFA	6.7	119.6	32.6	NA	NA	*23.2*	NA	NA	*44*	NA	NA	*23.4*	NA	NA	*10*	NA	NA	*57.6*	NA	NA	*41.6*
Validation experiment	6.7	120	32	59	45.5	*22.9*	12.6	7.2	*43.1*	61	48.2	*21*	22	20	*9.0*	15.8	6.3	*60.1*	1.6	0.8	*45.6*
Prediction error (%) for DFA						1.37			2.0			10.3			10			-4.4			-9.7
For straight bevel gear																					
RSM (Stage-3)	7	120	30	NA	NA	NA	NA	NA	NA	111	79.1	*28.7*	108.7	52.7	*51.5*	13	6.45	*50.4*	2.3	0.9	*62.1*
DFA	6.9	119.3	34.4	NA	NA	NA	NA	NA	NA	NA	NA	*28.7*	NA	NA	*53.9*	NA	NA	*49.6*	NA	NA	*64*
Validation experiment	6.9	120	34	NA	NA	NA	NA	NA	NA	72.2	53	*26.6*	68.7	33	*52.0*	13.6	7.4	*45.2*	2	0.7	*63*
Prediction error (%) for DFA						--			---			7.3			3.6			8.9			1.7

References

[1] Lee, D. H., I. J. Jeong and K. J. Kim. 2018. "A desirability function method for optimizing mean and variability of multiple responses using a posterior preference articulation approach." *Quality and Reliability Engineering International.* 34(3): 360-376. doi: 10.1002/qre.2258

[2] Gupta, K., N. K. Jain. 2014. "Analysis and optimization of the surface finish of the wire electrical discharge machined miniature gears." *Proceedings IMechE, Part B: Journal of Engineering Manufacture*. 228(5): 673-681. doi: 10.1177/0954405413508938

[3] Petare, A., N. K. Jain and I. A. Palani. 2021. "Simultaneous improvement of microgeometry and surface quality of spur and straight bevel gears by abrasive flow finishing process." *Journal of Micromanufacturing.* 4(2):189-206. doi:10.1177/25165984211021010

CHAPTER EIGHT

LASER TEXTURING OF THE GEARS

This chapter presents the results of the investigation of the role of laser texturing in enhancing the productivity and capabilities of the AFF process and to further improve microgeometry, surface finish, microhardness, surface morphology, and wear characteristics of spur and straight bevel gears.

8.1 Introduction to Gear Texturing

Past works on laser texturing (Sasi et al., 2017, Kang et al., 2017, Ye et al., 2018) have revealed that the most frequently used patterns for laser texturing are homothetic (parallel line), wavy (curved), spot, micro-dimples, and micro-pillars of square and triangular cross-sections. Homothetic texture has been shown to reduce the coefficient of friction by more than any other texture pattern (Hao et al., 2018). Therefore, a homothetic texture pattern was selected for generation on both the flank surfaces of all teeth of three spur and three straight bevel gears in a direction perpendicular to the lay pattern produced by the corresponding gear manufacturing process, as depicted in figs. 8.1a and 8.1b (subsequently these gears are referred to as laser textured gears). This resulted in the formation of a mesh-like structure between the homothetic laser textures and lay pattern on the flank surfaces of spur and straight bevel gear, as shown in figs. 8.1c and 8.1d. A fiber laser of 1064 nm wavelength generated by a computer-controlled machine ProMark (from *Scantech Laser Pvt. Ltd. Mumbai, India*; please refer to Appendix-B for its details) was used for laser texturing using the identified optimum values of laser power, focal length, and number of passes. Figure 8.2 depicts a photograph of this machine, which has a maximum power capacity of 50 W and the provision to vary the focal length, laser power, and number of passes.

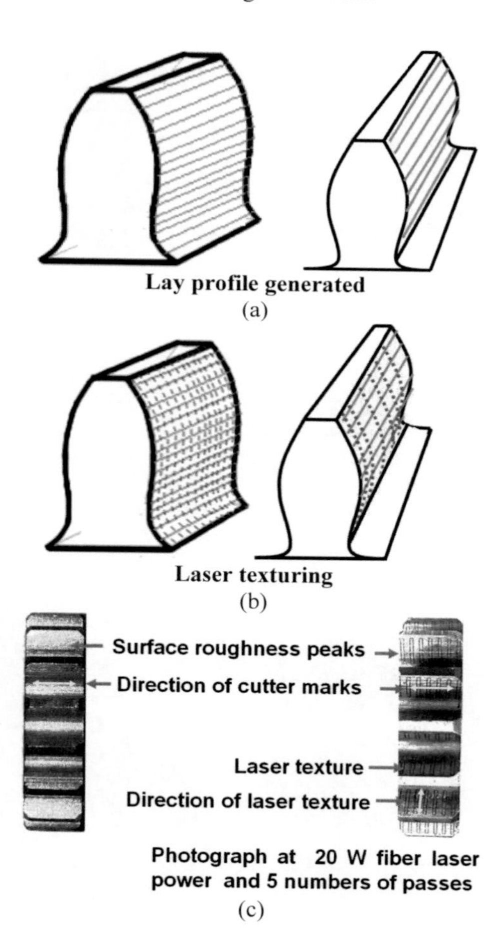
Lay profile generated
(a)
Laser texturing
(b)
Surface roughness peaks
Direction of cutter marks
Laser texture
Direction of laser texture
Photograph at 20 W fiber laser power and 5 numbers of passes
(c)

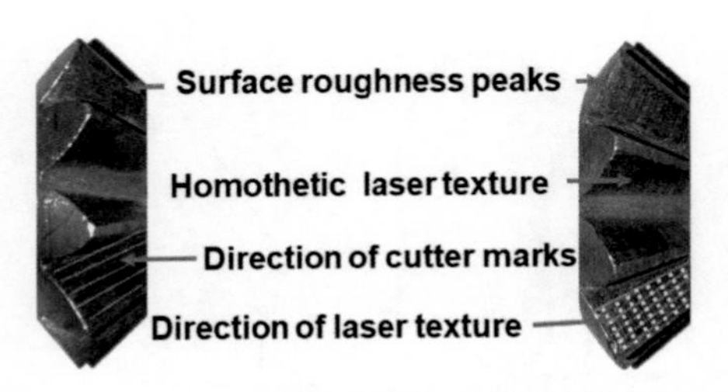

(d)

Fig. 8.1: Schematic of the lay profile of (a) spur and straight bevel gears, (b) laser-textured spur and straight bevel gears, (c) photograph of the untextured spur gear and laser textured spur gear using 20 W laser power and five passes, and (d) photograph of the untextured straight bevel gear and laser-textured straight bevel gear using 20 W laser power and five passes (Petare et al., 2018, Petare et al., 2020)

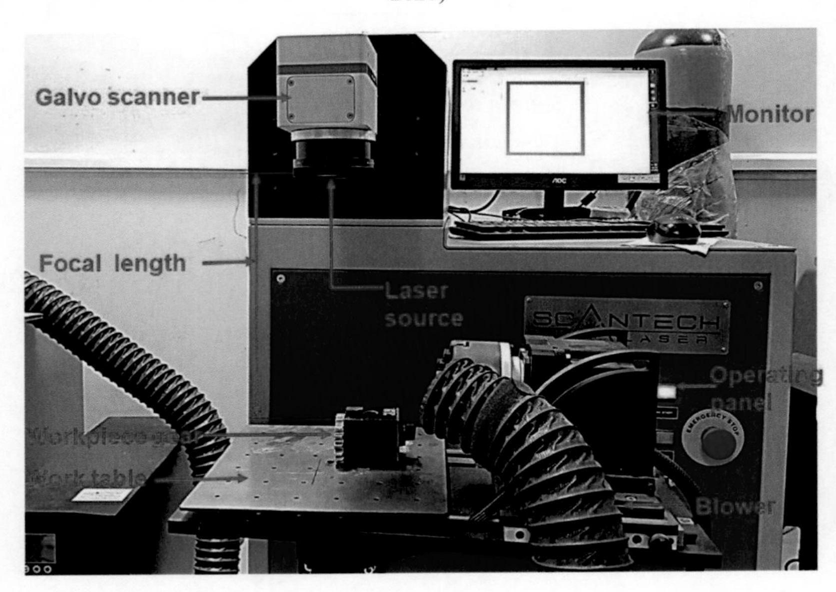

Fig. 8.2: Photograph of the continuous fiber laser machine used for the laser texturing of the workpiece gear

A total of 72 full factorial experiments were performed for the spur and straight gears each to identify the optimum values of laser power, focal length, and number of laser passes by varying laser power at four levels (i.e. 10, 15, 20, and 25 W), focal length at three levels (i.e. 280, 285, and 290 mm), and numbers of passes at six levels (i.e. 1, 2, 3, 4; 5, and 6) as presented in table 8.1. The optimum value of the focal length was identified according to the sharpness of the boundaries and corners of the marking area (product of marking height and length, also known as gain), which is determined according to the size of the flank surfaces of the gear tooth. A marking height of 2.2 mm and length of 4.4 mm were used. Figure 8.3 depicts the concept of the marking area for the focal length value of 280 mm (fig. 8.3a), 285 mm (fig. 8.3b), and 290 mm (fig. 8.3c) by a laser pointer. It is seen that a 285 mm focal length yielded very sharp boundaries and corners of the marking area. Therefore, it was identified as the optimum value of focal length. The formation of the laser textures on the flank surfaces of spur and straight bevel gears was checked using a 10x magnification lens. It was observed that no laser texture was formed for laser power less than 20 W and five passes. Fine laser textures were obtained for the combination of 20 W laser power and five passes. The use of 25 W laser power resulted in burn marks on the flank surfaces of the spur and straight bevel gears and increased density. Therefore, 20 W laser power and five passes are identified as the optimum values for laser texturing of spur and straight bevel gears. These values are presented in table 8.1

Table 8.1: Details of the fixed and variable parameters used to identify the optimum parameters for laser texturing of spur and straight bevel gears.

Variable parameters (unit)	***Level***					***Optimum value identified***
	I	II	III	IV	V	
Focal length (mm)	280	285	290	-	-	285
Laser power (watts)	10	15	20	25	-	20
Number of laser passes	1	2	3	4	5	5
Fixed parameters						
Laser wavelength (nm):1064; and Gain [marking length (mm) x height (mm)]: 1.9 [2.2 x 4.4]						

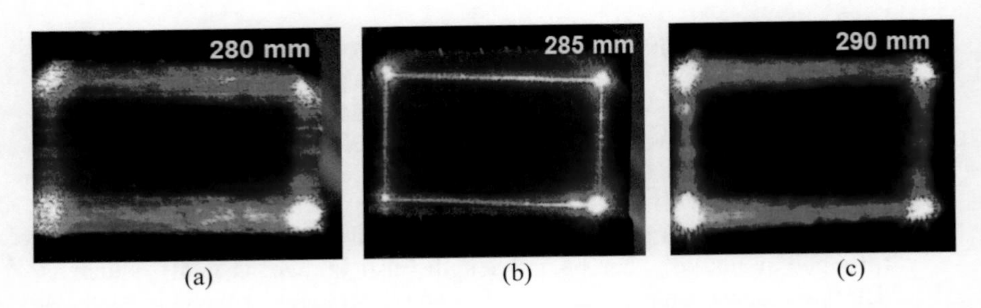

(a) (b) (c)

Fig. 8.3: Concept of boundaries of the marking area on the flank surface of gear for laser texturing at different values of focal length: (a) 280 mm, (b) 285 mm, and (c) 290 mm (Petare et al., 2018).

8.2 Comparison of the AFF-Finished Laser-Textured and Untextured Gears

Three untextured and three laser-textured spur gears and straight bevel gears (a total of six gears for each category) were used to perform experiments using the AFF process. The finishing time varied at three levels i.e. 15, 20, and 25 minutes for the spur gears, and 20, 25, and 30 minutes for the straight bevel gears using observations from the stage-2 experiments and 5 MPa extrusion pressure. The AFF medium had a silicone oil concentration of 10% vol. (with a corresponding medium viscosity of 135 kPa.s), which, as the stage-2 experiments identified, is optimum value, 100 mesh silicon carbide abrasive particle, a volumetric concentration of abrasive particles of 30%, and a 60% volumetric concentration of the putty (i.e. molding clay). Surface roughness and microgeometry parameters were used as the responses for finishing both the laser-textured and untextured gears. Surface roughness profile, surface morphology, wear-resistance, and microhardness were evaluated for the best finished laser-textured and the best finished untextured spur and straight bevel gears by the AFF process for the finishing duration of 25 minutes (for spur gears) and 30 minutes (for straight bevel gears). Volumetric material removal rate (MRR) was used to compare the productivity of the AFF process for finishing the untextured and laser-textured gears.

8.2.1 Comparison of Microgeometry Errors, Surface Roughness Parameters, and MRR

Table 8.2 presents the volumetric MRR and percentage reductions in microgeometry errors and surface roughness parameters for the AFF-finished untextured spur gear (USG), laser- textured spur gear (LTSG), untextured straight bevel gear (USBG), and laser-textured straight bevel gear (LTSBG) for three different values of the finishing time. Appendix C (iv) presents the charts for microgeometry errors given by the CNC Gear Metrology machine. Figure 8.4 depicts the effects of the finishing time for the AFF-finished untextured and laser-textured spur gear on avg. % reductions in total profile error 'PRF_a' and total lead error 'PRF_β' (fig. 8.4a); avg. % in total pitch error 'PRF_p' and % reduction in radial runout 'PRF_r' (fig. 8.4b); and avg. % in max. and avg. surface roughness 'PRR_{max}' and PRR_a' (fig. 8.4c). Similarly, fig. 8.5 depicts the effects of the finishing time for the AFF-finished untextured and laser- textured straight bevel gear on avg. % reduction in total pitch error 'PRF_p' and % reduction in radial runout 'PRF_r' (fig. 8.5a); and avg.% reductions in max. and avg. surface roughness 'PRR_{max}' and 'PRR_a' (fig. 8.5b). Figure 8.6 depicts the effects of the finishing time on MRR of the AFF-finished untextured and laser-textured spur gear (fig. 8.6a) and the AFF- finished untextured and laser-textured straight bevel gear are depicted in fig. 8.6b.

Table 8.2: Values of MRR, % reductions in microgeometry errors, and surface roughness parameters for the AFF finished untextured and laser textured spur and straight bevel gears for different values of finishing time (Petare et al., 2018, Petare et al., 2020).

Exp. No.	Finishing time (min)	Gear type	*Avg % reductions in (except in radial runout)* Total profile error 'PRF_a'	Total lead error 'PRF_β'	Total pitch error 'PRF_P'	Radial runout 'PRF_r'	Avg. surface roughness 'PRR_a'	Max. surface roughness 'PRR_{max}'	MRR (mm^3/min)
			For ***spur*** gear						
1	15	USG	6.9	8.5	2.1	1.2	39.8	26.5	1.9
		LTSG	9.3	11.7	6.1	1.9	46.4	36.8	3.9
2	20	USG	10.4	11.9	14.2	2.9	49.8	52.8	4.0
		LTSG	16.6	27.5	15.6	3.4	51.2	44.6	6.3
3	25 (Change in DIN quality)	USG	20.6 11→10	38.9 8→6	15.7 11→10	4.2 No change	66.2	57.2	6.4
		LTSG	28.5 12→11	40.2 11→9	24.9 11→10	4.8 No change	72.4	68.9	8.5
			For ***straight bevel*** gear						
1	20	USBG	NA	NA	24.2	42.0	47.8	53.1	4.1
		LTSBG			25.3	46.8	48.0	53.9	5.0
2	25	USBG	NA	NA	27.3	67.7	52.1	56.3	6.5
		LTSBG			28.0	70.9	53.4	57.2	7.6
3	30 (Change in DIN quality)	USBG	NA	NA	29.6 11→10	47.8 11→9	58.4	62.1	8.1
		LTSBG			32.3 10→8	48.5 9→6	60.5	62.8	8.7

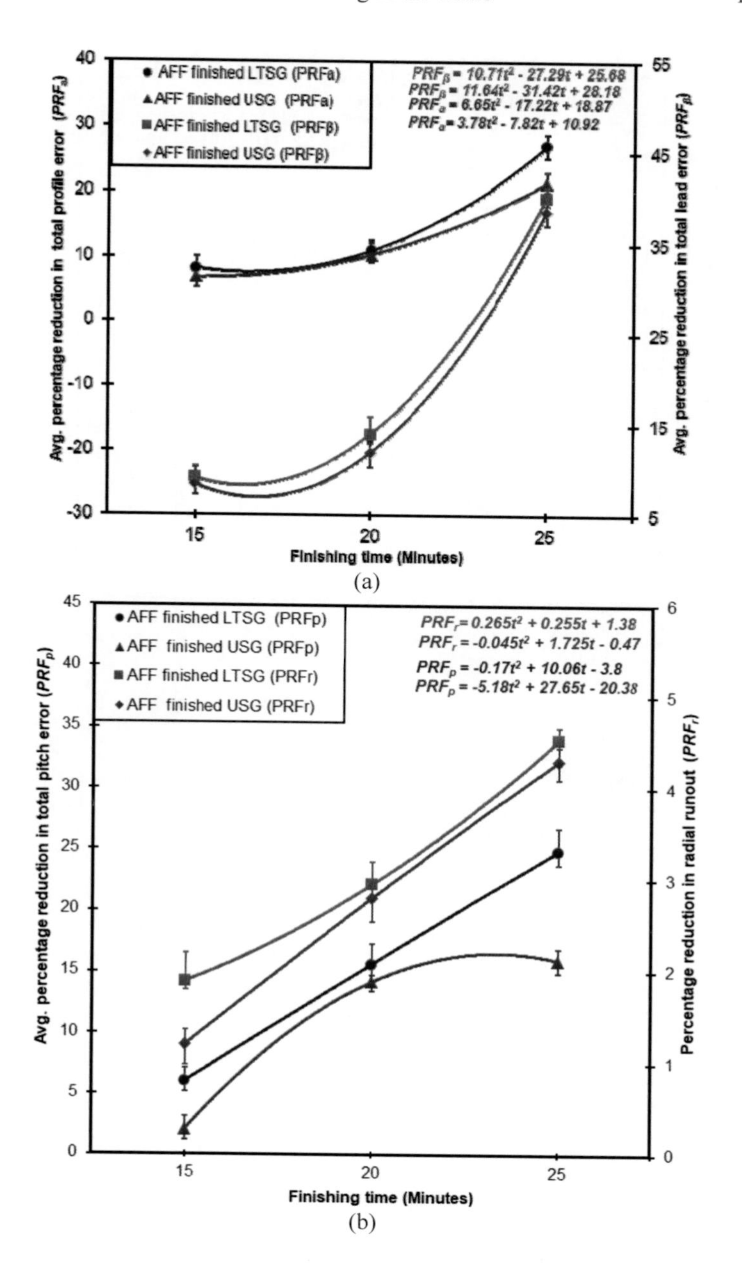
AFF finished LTSG (PRFa)
AFF finished USG (PRFa)
AFF finished LTSG (PRFβ)
AFF finished USG (PRFβ)
PRFβ = 10.71t² - 27.29t + 25.68
PRFβ = 11.64t² - 31.42t + 28.18
PRFα = 6.65t² - 17.22t + 18.87
PRFα = 3.78t² - 7.82t + 10.92
Avg. percentage reduction in total profile error (PRFα)
Avg. percentage reduction in total lead error (PRFβ)
Finishing time (Minutes)
AFF finished LTSG (PRFp)
AFF finished USG (PRFp)
AFF finished LTSG (PRFr)
AFF finished USG (PRFr)
PRFr = 0.265t² + 0.255t + 1.38
PRFr = -0.045t² + 1.725t - 0.47
PRFp = -0.17t² + 10.06t - 3.8
PRFp = -5.18t² + 27.65t - 20.38
Avg. percentage reduction in total pitch error (PRFp)
Percentage reduction in radial runout (PRFr)
Finishing time (Minutes)

(a)

(b)

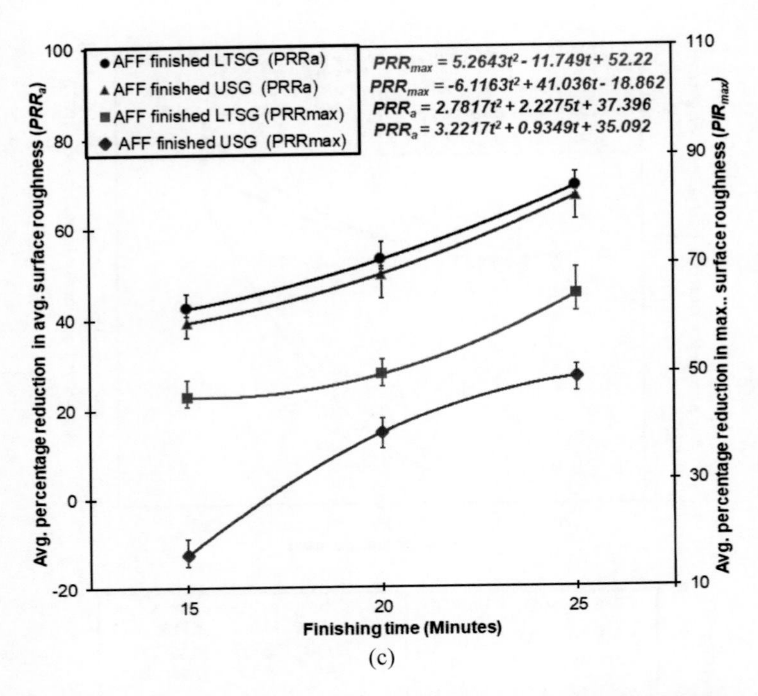

(c)

Fig. 8.4: Effect of the finishing time on % reductions in (a) avg. total profile error 'PRF_a' and avg. total lead error 'PRF_β', (b) avg. total pitch error 'PRF_p' and radial runout 'PRF_r', and (c) max surface roughness 'PRR_{max}'; and avg. surface roughness 'PRR_a' for the AFF finished untextured spur gear (USG) and laser textured spur gear (LTSG) (Petare et al., 2018)

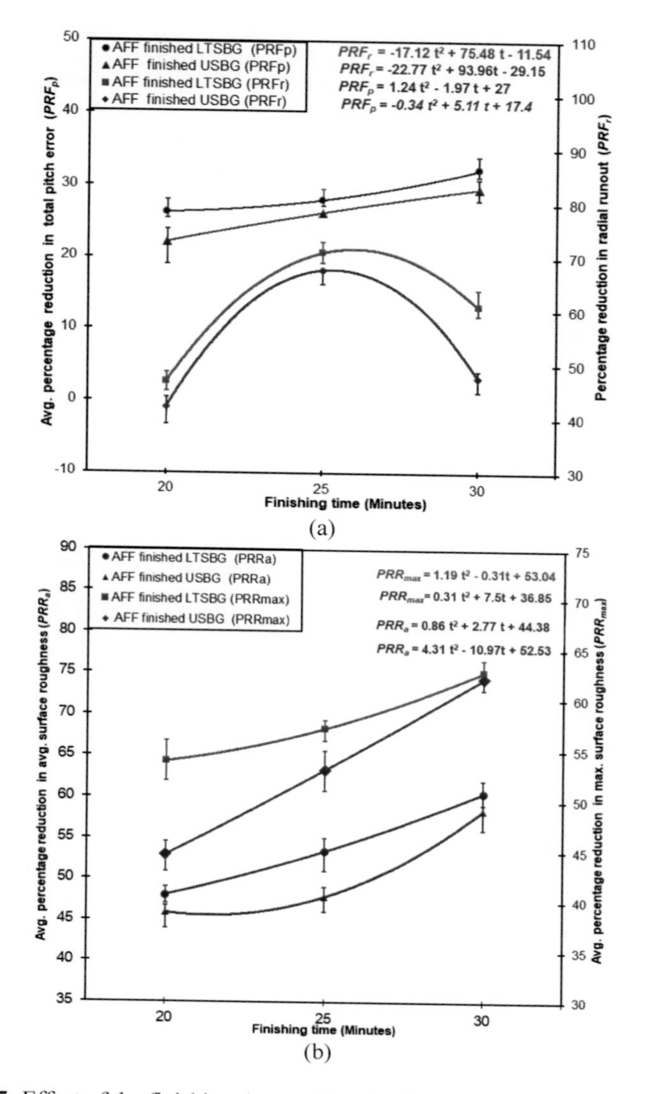

Fig. 8.5: Effect of the finishing time on % reductions in (a) avg. total pitch error 'PRF_p' and radial runout 'PRF_r', and (b) max. surface roughness 'PRR_{max}'; and avg. surface roughness 'PRR_a' for the AFF-finished untextured straight bevel gear (USBG) and laser-textured straight bevel gear (LTSBG) (Petare et al., 2020)

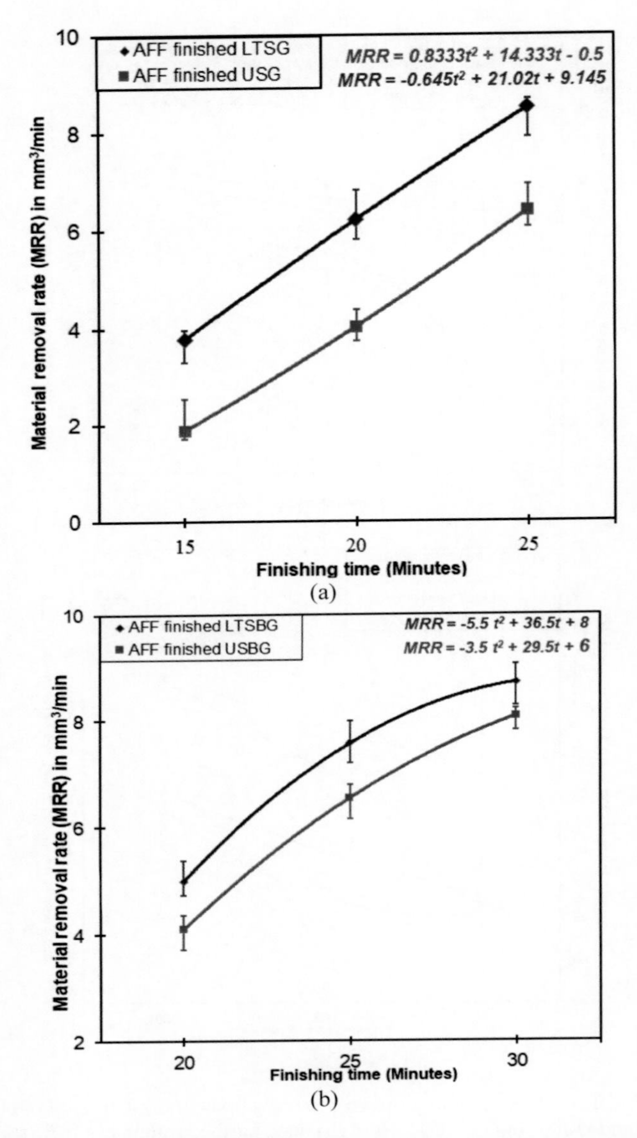

Fig. 8.6: Variation of material removal rate with the finishing time for the (a) untextured and laser textured spur gear, and (b) untextured and laser textured straight bevel gear (Petare et al., 2018, Petare et al., 2020)

The following observations can be made from the results of table 8.2 and the graphs presented in figs. 8.4 to 8.6:

- An increase in the AFF finishing time monotonically increases vol. MRR, avg. percentage reductions in total profile error (PRF_a), total lead error (PRF_{β},), total pitch error (PRF_p), max. surface roughness (PRR_{max}) and avg. surface roughness (PRR_a), and percentage reduction in radial runout (PRF_r) of both the untextured and laser-textured ***spur*** gears, and attain their maximum values at 25 minutes of finishing time. Similarly, vol. MRR, avg. percentage reductions in total pitch error (PRF_p), max. surface roughness (PRR_{max}), and avg. surface roughness (PRR_a) of both the untextured and laser-textured ***straight bevel*** gears increase monotonically with the attainment of their maximum values at 30 minutes of finishing time.
- Percentage reduction in radial runout 'PRF_r' of the untextured and laser-textured straight bevel gear shows the optimum finishing time of 25 minutes (fig. 8.5a) because after increasing up to 25 minutes, it starts decreasing thereafter. This is due to the wear of the abrasive particles and the decrease in viscosity of the AFF medium caused by the mixing of swarf (i.e. removed material during AFF of the gears), which causes a scratching and rubbing of the finished flank surfaces of the gear, thus deteriorating their microgeometry and surface roughness.
- The AFF of the laser-textured spur and straight bevel gears gives higher values of vol. MRR and higher percentage reductions in their microgeometry errors and surface roughness parameters than the AFF of the untextured spur and straight bevel gears for all the finishing times, and the differences between them widen with increases in the finishing time and become maximum at its highest value. This can be explained with the help of the mechanism of finishing the untextured and laser-textured spur and straight bevel gears, as depicted in fig. 8.7. Fig. 8.7 shows the interaction pattern of the active abrasive particles in the AFF process with the untextured and laser-textured flank surfaces. During the AFF of the untextured flank surfaces, the abrasive particles follow a straight path along the lay pattern generated by the gear cutter (fig. 8.7a) and travel less distance with less active abrasive particles, whereas in the case of laser-textured flank surfaces, a mesh-like structure is formed between the homothetic laser textures and lay pattern of the gear cutter (fig. 8.7b). It deflects the flow of the abrasive particles during the AFF process and provides them with a random motion

with more active abrasive particles, thus enabling them to cover a longer flow path over the flank surfaces. It results in more uniform abrasion, more MRR, and more reductions in the considered responses in less finishing time.

- Improvements in the considered responses are more pronounced in case of the AFF of laser- textured *spur* gears than in the laser-textured *straight bevel* gears. This is due to the higher hardness of the bevel gear material, which results in the removal of less material and fewer reductions in microgeometry errors and surface roughness.
- AFF improved the (i) quality of the untextured spur gear from DIN 11 to DIN 10 and the laser-textured spur gear from DIN 12 to DIN 10 for total profile error; (ii) quality of the untextured spur gear from DIN 8 to DIN 6 and the laser-textured spur gear from DIN 10 to DIN 8 for total lead error; (iii) quality of both the untextured and laser-textured spur gears from DIN 11 to DIN 10 for total pitch error; (iv) no change in the quality of both the untextured and laser-textured spur gears for radial runout, although there is a slight reduction in their values after the AFF; (v) quality of the untextured straight bevel gear from DIN 11 to DIN 10 and of laser-textured straight bevel gear from DIN 10 to DIN 8 for total pitch error; and (vi) quality from the untextured straight bevel gear from DIN 11 to DIN 9 and that of the laser-textured straight bevel gear from DIN 9 to DIN 6 for radial runout.

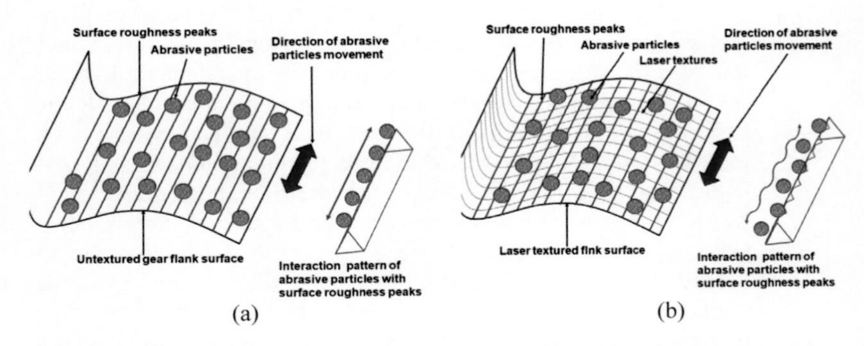

Fig. 8.7: Interaction pattern of the active abrasive particles with flank surfaces of the (a) untextured gear, and (b) laser textured gear (Petare et al., 2020)

8.2.2 Comparison of Surface Roughness Profiles

Figures 8.8 and 8.9 present the profile of the surface roughness for the (i) untextured spur gear (fig. 8.8a), AFF-finished untextured spur gear (fig. 8.8b), laser-textured spur gear (fig. 8.8c), and AFF-finished laser-textured spur gear (fig. 8.8d); and (ii) untextured straight bevel gear (fig. 8.9a), AFF-finished untextured straight bevel gear (fig. 8.9b), laser-textured straight bevel gear (fig. 8.9c), and AFF-finished laser-textured straight bevel gear (fig. 8.9d). It can be noticed from these figures that the AFF of the untextured spur gear reduced max. and avg. surface roughness values from 18.3 to 7.7 μm and 1.9 to 0.6 μm, respectively, and the AFF of the laser-textured spur gear reduced their values from 12.9 to 4 μm and 2 to 0.5 μm, respectively. Similarly, the AFF of the untextured straight bevel gear reduced max. and avg. surface roughness from 14 to 5.3 μm and 1.6 to 0.7 μm, respectively, whereas the AFF of the laser-textured straight bevel gear reduced their values from 16.5 to 6.1 μm to 2.2 to 0.8 μm, respectively, i.e. % reductions in max. and avg. surface roughness values of the laser-textured spur and straight bevel gear exceed those in the corresponding untextured gears.

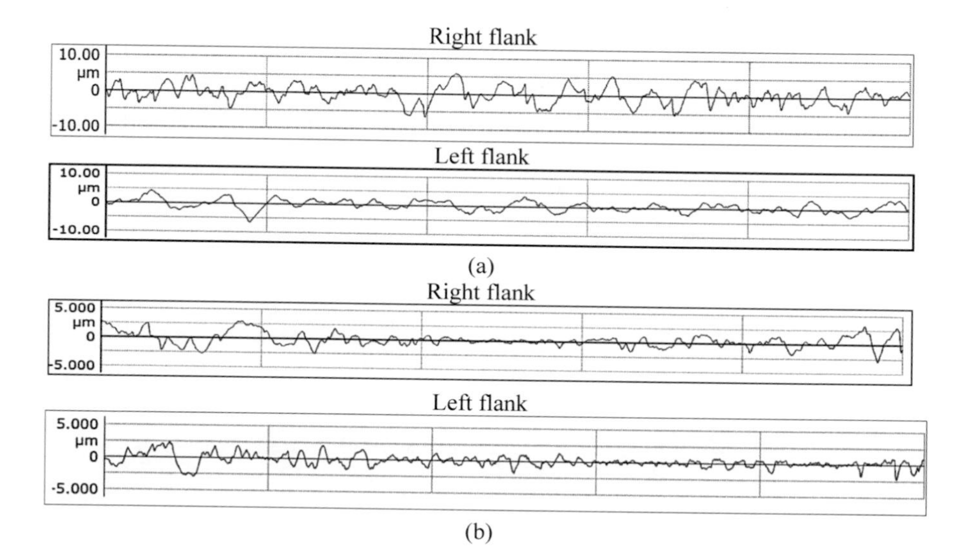

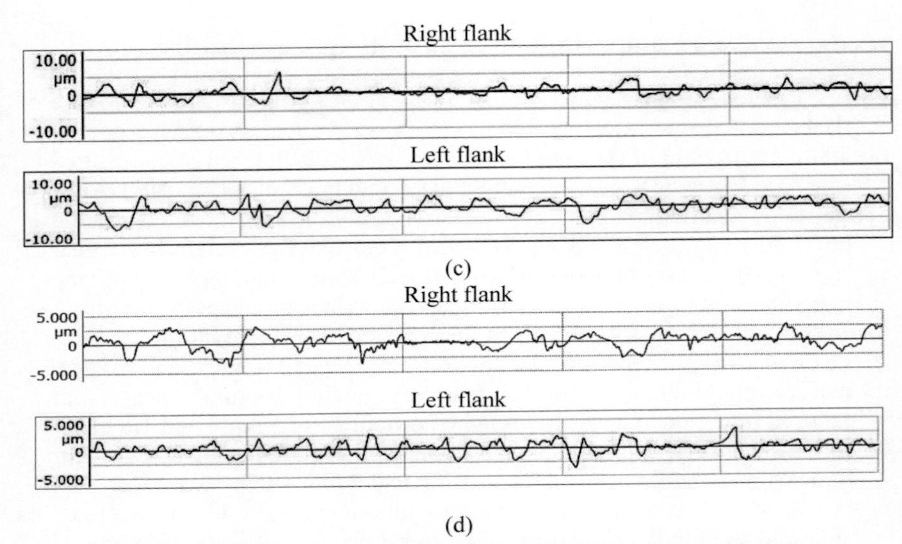

Fig. 8.8: Surface roughness profiles of the right and left flanks of the (a) untextured spur gear, (b) AFF-finished untextured spur gear, (c) laser-textured spur gear, and (d) AFF-finished laser-textured spur gear (Petare et al., 2018)

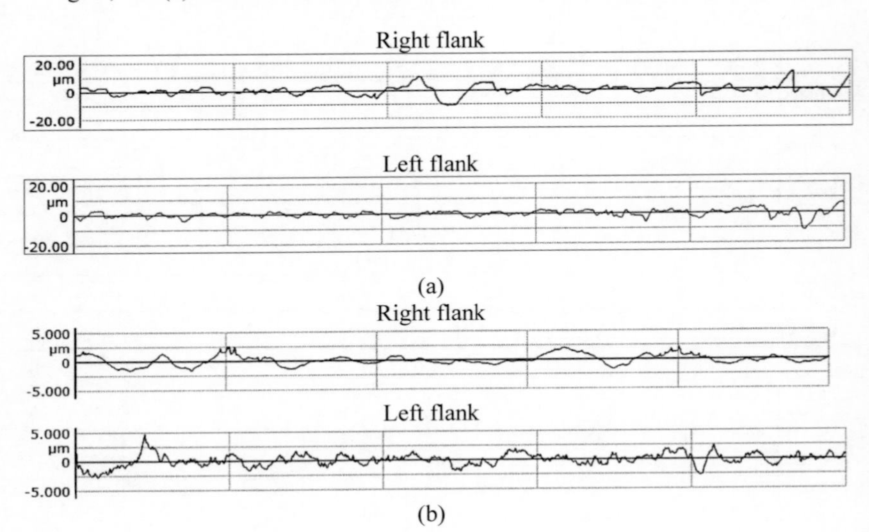

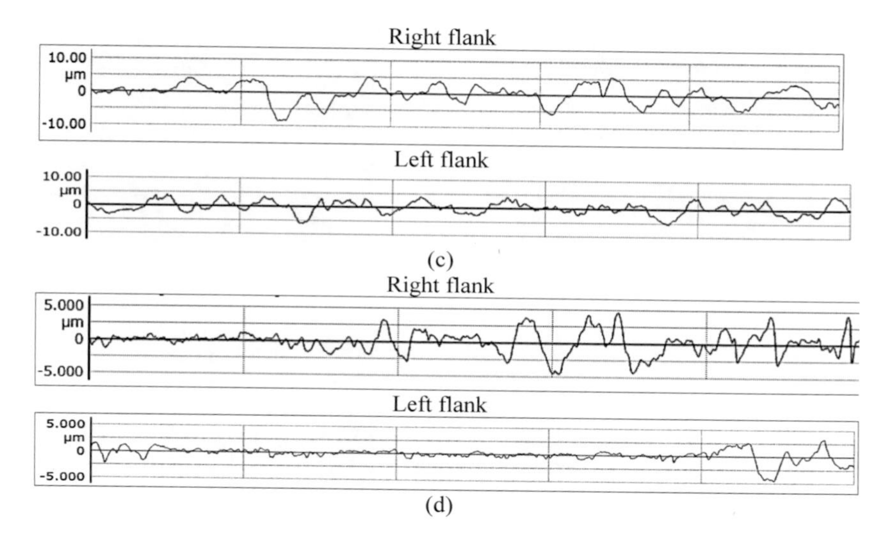

Fig. 8.9: Surface roughness profiles of the right and left flanks of the (a) untextured straight bevel gear, (b) AFF- finished untextured straight bevel gear, (c) laser-textured straight bevel gear, and (d) AFF-finished laser- textured straight bevel gear (Petare et al., 2020)

8.2.3 Comparison of Surface Morphology

Figures 8.10 and 8.11 depict SEM micrographs showing the surface morphology of the flank surfaces of (i) untextured (fig. 8.10a), AFF-finished untextured (fig. 8.10b), and AFF-finished laser-textured spur gears (fig. 8.10c); and (ii) untextured (fig. 8.11a), AFF-finished untextured (fig. 8.11b), and AFF-finished laser-textured straight bevel gears (fig. 8.11c). It can be seen from these images that the AFF of the untextured spur (fig. 8.10b) and straight bevel gear (fig. 8.11b) completely removes burrs, roughness peaks, and cutter marks from their corresponding unfinished flank surfaces (figs. 8.10a and 8.11a), giving a smooth surface with the presence of microchips. Comparing SEM micrographs of the AFF-finished untextured (figs. 8.10b and 8.11b) and AFF-finished laser-textured spur and straight bevel gears (figs. 8.10c and 8.11c) makes it clear that the flank surfaces of the AFF-finished laser textured gears are smoother than those of the AFF-finished untextured gears due to the reduced number of microchips and pealed material.

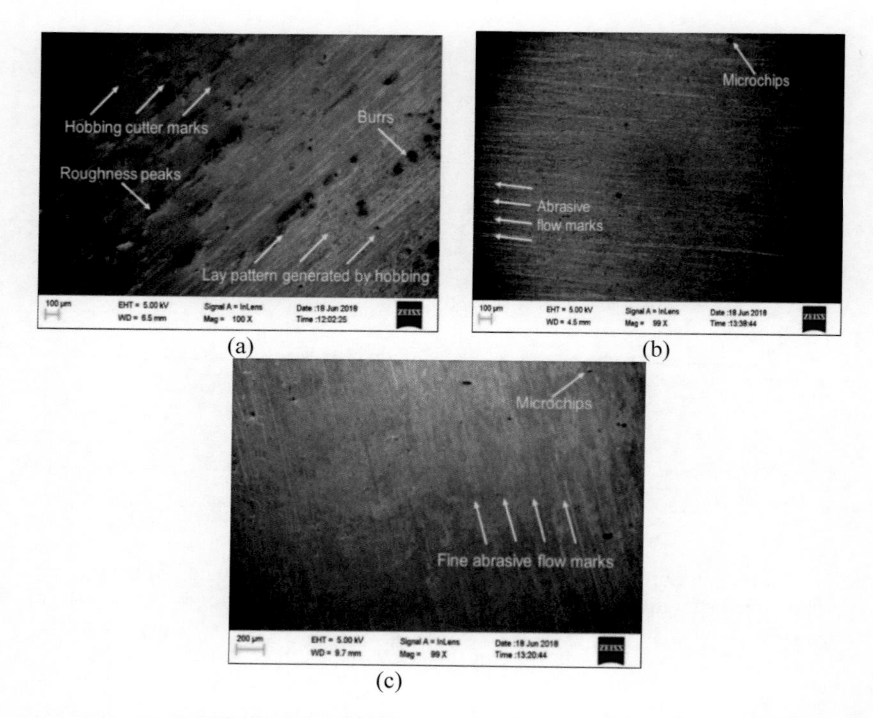

Fig. 8.10: SEM images depicting surface morphology of the flank surface of the (a) untextured spur gear, (b) AFF-finished untextured spur gear, and (c) AFF-finished laser-textured spur gear (Petare et al., 2018)

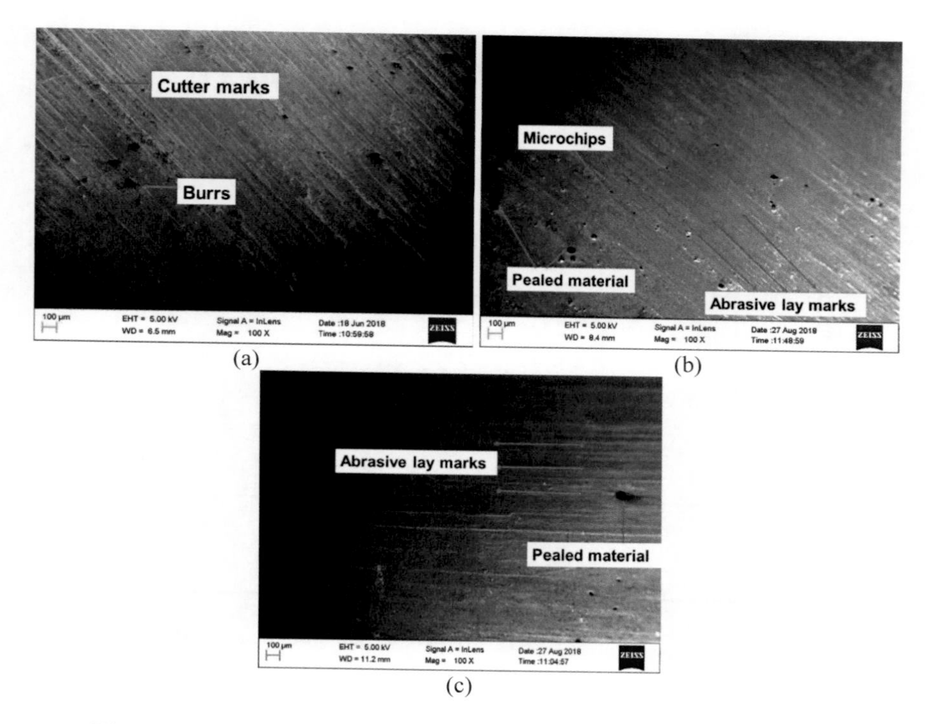

(a) (b)

(c)

Fig. 8.11: SEM images showing surface morphology of the flank surface of (a) untextured straight bevel gear, (b) AFF-finished untextured straight bevel gear, and (c) AFF-finished laser-textured straight bevel gear (Petare et al., 2018)

8.2.4 Comparison of Wear Characteristics

Figures 8.12 and 8.13 depict variations of the coefficient of friction and the frictional forces with fretting wear time for the untextured, AFF-finished untextured, and AFF-finished laser- textured spur gears (gig. 8.12), and untextured, AFF-finished untextured, and AFF-finished laser-textured straight bevel gear (fig. 8.13). Values of the wear characteristics (i.e. frictional force and the coefficient of friction, specific wear rate, wear rate and wear volume) obtained from the graphs are presented in table 8.3. Figure 8.14 depicts the SEM micrograph showing the surface morphology of worn flank surfaces of the AFF-finished (i) untextured spur (fig. 8.14a) and straight bevel gears (fig. 8.14b); and (ii) laser textured spur (fig. 8.14c) and straight bevel gears (fig. 8.14d).

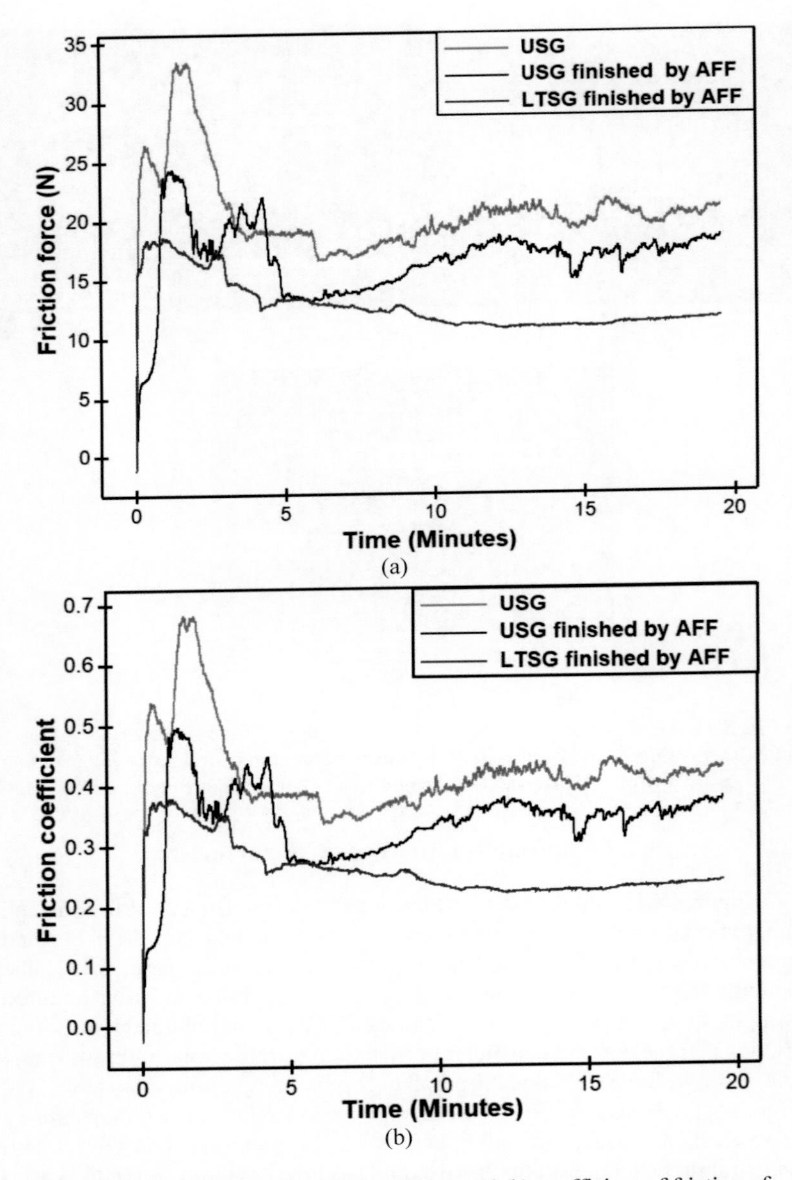

Fig. 8.12: Variation of the (a) friction force; and (b) coefficient of friction of untextured, AFF-finished untextured, and AFF-finished laser-textured spur gears with fretting wear time (Petare et al., 2018)

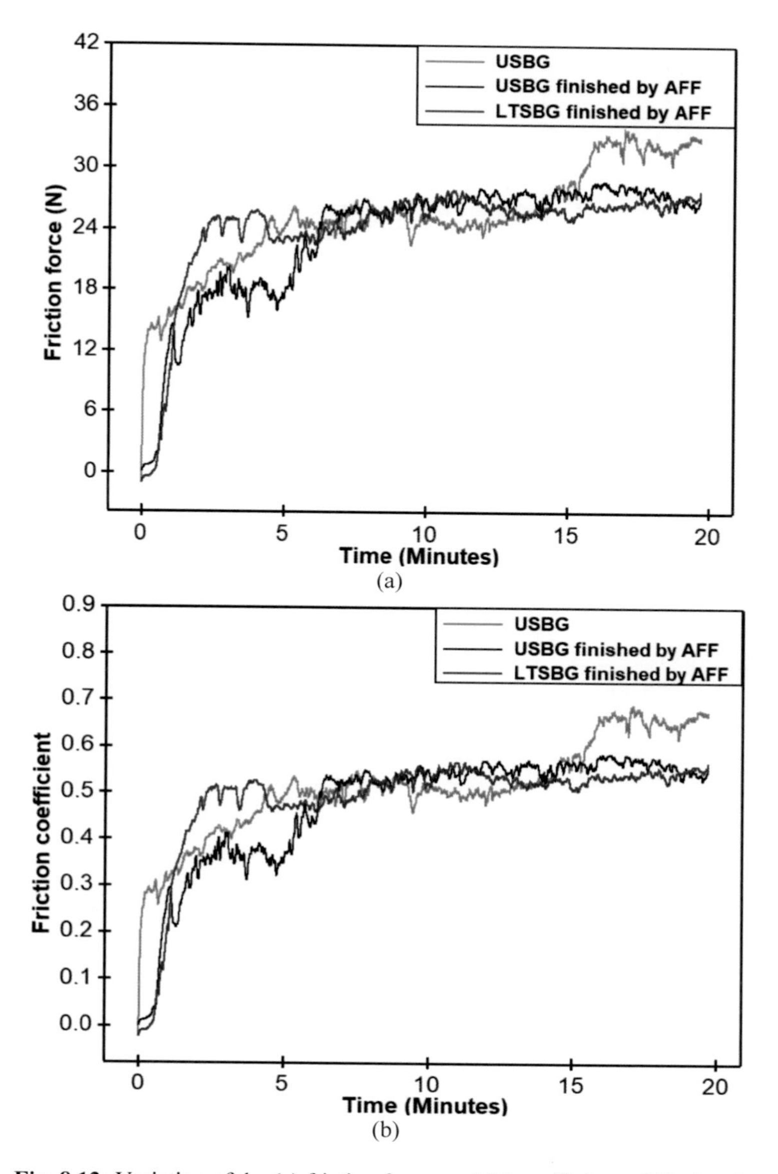

Fig. 8.13: Variation of the (a) friction force; and (b) coefficient of friction for untextured, AFF-finished untextured, and AFF-finished laser-textured straight bevel gears with fretting wear time (Petare et al., 2020)

Table 8.3: Results of fretting wear test for the untextured, AFF finished untextured and AFF finished laser textured spur and straight bevel gears (Petare et al., 2018, Petare et al., 2020)

Wear characteristics (unit)	*Spur gear*			*Straight bevel gear*		
	Untextured	AFF-finished untextured	AFF-finished laser-textured	Untextured	AFF-finished untextured	AFF-finished laser-textured
Max. value of frictional force (N)	33.4	24.3	18.6	33.9	28.8	27.9
Max. value of coefficient of friction	0.7	0.5	0.3	0.7	0.58	0.52
Specific wear rate '*k*' (mm^3/N-m)	14.8×10^{-6}	5.7×10^{-6}	4.2×10^{-6}	9×10^{-6}	5.9×10^{-6}	6×10^{-6}
Wear rate (mm^3/m)	7.4×10^{-4}	2.8×10^{-4}	2.1×10^{-4}	4.4×10^{-4}	2.99×10^{-4}	2.97×10^{-4}
Wear volume '*V*' (mm^3)	0.17	0.06	0.04	0.96	0.71	0.70

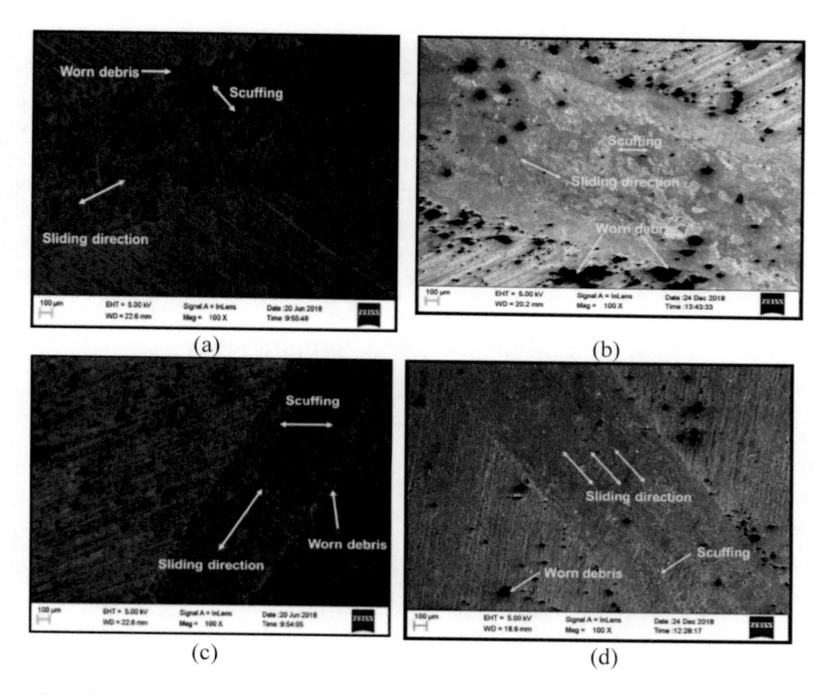

Fig. 8.14: Surface morphology of the worn flank surface of the AFF finished untextured (a) spur gear and (b) straight bevel gear, and AFF finished laser textured (c) spur gear, and (d) straight bevel gear (Petare et al., 2018)

The following interferences can be drawn from figs. 8.12–8.14 and table 8.3.

- The untextured spur and straight bevel gears have maximum values of all the wear characteristics, and the AFF-finished laser-textured spur and straight bevel gears have their minimum values. Values for the AFF-finished untextured gears lie between them. This implies that (i) laser texturing helps in further reduction of all wear characteristics. This is due to the improved microhardness of the flank surfaces caused by a distortion of the lattice structure by laser texturing, which forms compressive residual stress (Dai et al., 2018), and (ii) the AFF decreases all the wear characteristics of both untextured and laser- textured gears due to the production of smoother flank surfaces with less variation in the surface roughness peaks by the active abrasive particles contained in the

AFF medium. Such flank surfaces offer more contact area, which increases their resistance to the applied load on the steel ball during the fretting wear test. This results in less breakage of the surface peaks and a higher resistance to wear compared to the untextured gears.

- Comparison of the surface morphology images in fig. 8.14 reveals the least worn debris, with much less pilling of the displaced material at the edges of the wear track in the AFF-finished laser-textured gears than that of the AFF-finished untextured gears.

8.2.5 Microhardness comparison

Figure 8.15 depicts the results of the microhardness evaluation for the untextured, the AFF- finished untextured, and the AFF-finished laser-textured spur gear (fig. 8.15a) and straight bevel gear (fig. 8.15b). It can be seen from these figures that:

- The AFF-finished laser-textured gears have a maximum value of microhardness and untextured gears have a minimum value of microhardness, and the microhardness of the AFF-finished untextured gears lies in between them. This is due to high-density displacement by the laser creating a structure of sub-grain boundary that blocks the plastic flow of the material, thus increasing the microhardness of the laser- textured surface (Dai et al., 2018).
- Extrusion pressure in the AFF process exerts axial and radial forces on the abrasive particles during the back-and-forth motion of the AFF medium through the restriction between fixture and workpiece. The continuous impact of the abrasive particles removes burrs, surface roughness peaks, and surface hardening of the flank surfaces. It increases the microhardness of the AFF-finished components. It also develops compressive residual stress, which improves fatigue strength (Sankar et al., 2009).

The above-mentioned phenomena are responsible for increasing the microhardness of the AFF-finished laser-textured gears more than that of the AFF- finished untextured gears, whereas only the last phenomenon (i.e. Exertion of axial and radial forces on the abrasive particles by extrusion pressure) is responsible for increasing the microhardness of the AFF finished untextured gears.

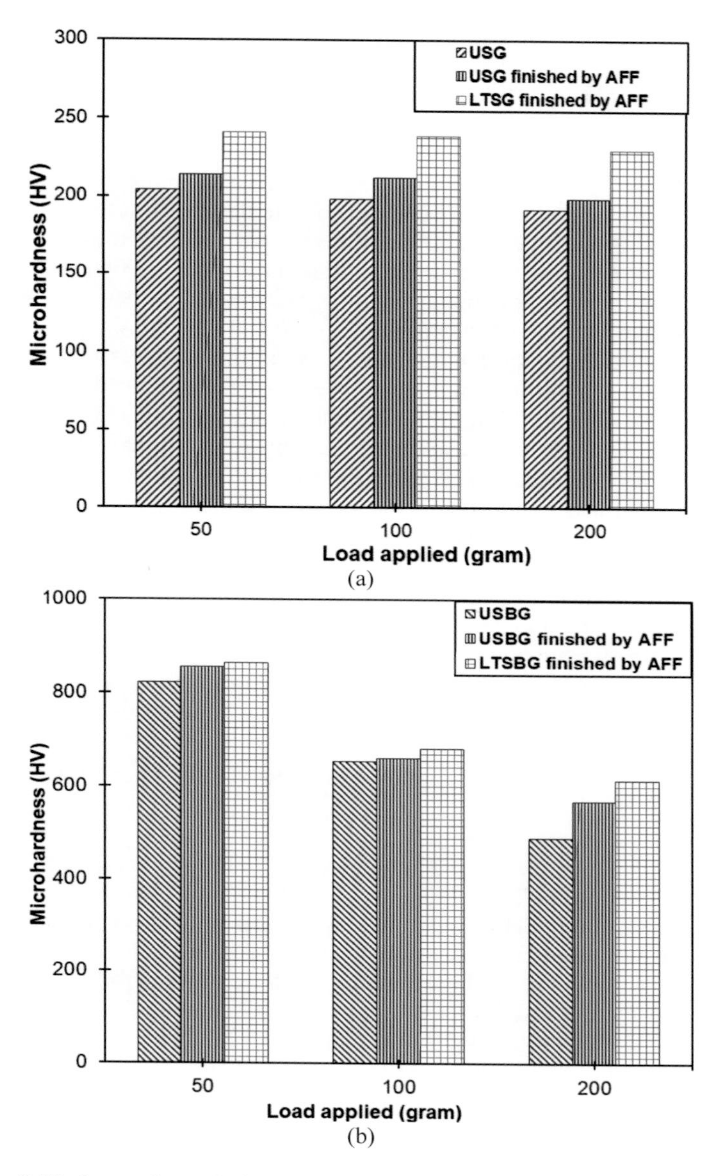

Fig. 8.15: Comparison of microhardness values for the untextured, AFF-finished untextured and AFF-finished laser-textured (a) spur gears; and (b) straight bevel gears (Petare et al., 2018, Petare et al., 2020)

8.3 Concluding Remarks

The following conclusions can be drawn from the investigations on laser texturing of spur and straight bevel gears:

- Laser power, focal length, and number of passes significantly influence the homothetic texture on the flank surfaces of gears in a direction perpendicular to the lay pattern generated by gear manufacturing process. A combination of 20 W of fiber laser, 285 mm focal length, and five passes were found to be the optimum values for the present study.
- The surface morphology study of the flank surfaces of the AFF-finished laser-textured gears reveals a smoother surface than the flank surfaces of the AFF-finished untextured gears.
- The AFF yields more microhardness in the laser-textured gears than in the untextured gears due to the continuous impact of the abrasive particles caused by extrusion pressure and the formation of subgrain boundaries by laser texturing.
- The AFF-finished laser-textured gears offer higher wear resistance compared to the untextured gears due to a smoother surface with less variation in the roughness profile produced by the AFF process and distortion of the lattice structure produced by the laser texturing, which forms compressive residual stresses.
- Laser texturing is an effective process to enhance the productivity of the AFF process in terms of improvement in microgeometry, surface finish, surface morphology, wear resistance, microhardness, and MRR.

References

[1] Dai, F. Z., J. Geng, W. S. Tan, X. D. Ren, J. Z. Lu and S. Huang. 2018. "Friction and wear on laser textured Ti6Al4V surface subjected to laser shock peening with contacting foil." *Optics & Laser Technology.* 103: 142-150. doi:10.1016/j.optlastec.2017.12.044

[2] Hao, X., X. Chen, S. Xiao, L. Li and N. He. 2018. "Cutting performance of carbide tools with hybrid texture." *The International Journal of Advanced Manufacturing Technology.* 97(9): 3547-3556. doi:10.1007/s00170-018-2188-2

[3] Kang, Z., Y. Fu, J. Ji and J. C. Puoza. 2017. "Effect of local laser surface texturing on tribological performance of injection cam." *The*

International Journal of Advanced Manufacturing Technology. 92(5): 1751-1760. doi:10.1007/s00170-017-0227-z

[4] Petare, A. C., A. Mishra, I. A. Palani and N. K. Jain. 2018. "Study of laser texturing assisted abrasive flow finishing for enhancing surface quality and microgeometry of spur gears." *The International Journal of Advanced Manufacturing Technology*.101(1): 785-799. doi: 10.1007/s00170-018-2944-3

[5] Petare, A., G. Deshwal, I. A. Palani and N. K. Jain. 2020. "Laser texturing of helical and straight bevel gears to enhance finishing performance of AFF process." *The International Journal of Advanced Manufacturing Technology*. 110(7): 2221-2238. doi:10.1007/s00170-020-06007-0

[6] Sankar, M. R., V. K. Jain and J. Ramkumar. 2009. "Experimental investigations into rotating workpiece abrasive flow finishing." *Wear*. 267(1): 43-51. doi:10.1016/j.wear.2008.11.007

[7] Sasi, R., S. K. Subbu and I. A. Palani. 2017. "Performance of laser surface textured high speed steel cutting tool in machining of Al7075-T6 aerospace alloy." *Surface and Coatings Technology*. 313: 337-346. doi:10.1016/j.surfcoat.2017.01.118

[8] Ye, D., Y. Lijun, C. Bai, W. Xiaoli, W. Yang and X. Hui. 2018. "Investigations on femtosecond laser-modified microgroove-textured cemented carbide YT15 turning tool with promotion in cutting performance." *The International Journal of Advanced Manufacturing Technology*. 96(9): 4367-4379. doi:10.1007/s00170-018-1906-0

CHAPTER NINE

STUDY OF THE PERFORMANCE CHARACTERISTICS OF GEARS

This chapter presents the results and analyses of the performance characteristics of the gears studied in terms of functional performance parameters determined by dual flank roll testing and noise and vibrations of the unfinished and the best finished spur and straight bevel gears from the stage-3 experimentation.

9.1 Evaluation of Noise and Vibrations of Gears

The unfinished and the best finished spur and straight bevel gears by the AFF process in the stage-3 experimentation (i.e. finished using the identified optimum values of extrusion pressure at 7 MPa; vol. concentration and size of SiC abrasive particles at 30% and 120 mesh; vol. concentration of silicone oil at 10%; finishing time at 25 minutes for spur gear and 30 minutes for straight bevel gears) were used for their noise and vibration testing on an in-house developed experimental test rig, depicted in fig. 9.1a. It has two independent motor-driven separate gearboxes: (i) one developed by Kumar et al. (2017) for testing the workpiece (or test) spur gear and its master gear mounted on the parallel shafts (fig. 9.1b) and supported by pedestal bearings at the other end, and (ii) another one developed by Kashyap et al. (2017) for testing the straight bevel gear and its master gear mounted on two perpendicular shafts (fig. 9.1c) and supported by pedestal bearings at the other end. The noise and vibration of the unfinished and the stage-3 experimentation best finished spur gear and straight bevel gears were measured by varying the speed and the applied load at four levels each (i.e. conducting 16 experiments for each type of gear), as presented in table 9.1. The motor of a gearbox was run at the desired speed. The desired value of the load was applied manually on the driven spur or straight bevel gear. A microphone (placed at the standard distance of 1 meter from the concerned gearbox) and a tri-axial accelerometer (mounted over the concerned gearbox) were used to record the noise and vibration, respectively, for the gear under investigation. These signals were transferred to the four-channel noise and vibrations data acquisition system OR 35 (from *OROS, France,* please refer to Appendix-B for its details). Its associated software (NV Gate 9.0, 3-series) was used to analyze the acquired signals. Changes in the noise level '*ΔN*' and vibration level '*ΔV*' were computed using the following equations:

$$\Delta N\ (dBA) = Noise\ of\ a\ gear\ before\ AFF - Noise\ of\ the\ same\ after\ AFF \quad (9.1)$$

$$\Delta V \left(\frac{m}{s^2}\right) = Vibration\ of\ a\ gear\ before\ AFF - Vibration\ of\ the\ same\ after\ AFF \quad (9.2)$$

Table 9.1: Details of parameters used in the measurement of noise and vibration of the unfinished and the Stage-3 experimentation best finished spur and straight bevel gears.

Process parameters	Levels and values			
Variable parameters	I	II	III	IV
Rotational speed (rpm)	250	500	750	1000
Applied load (N)	3	5	7	9

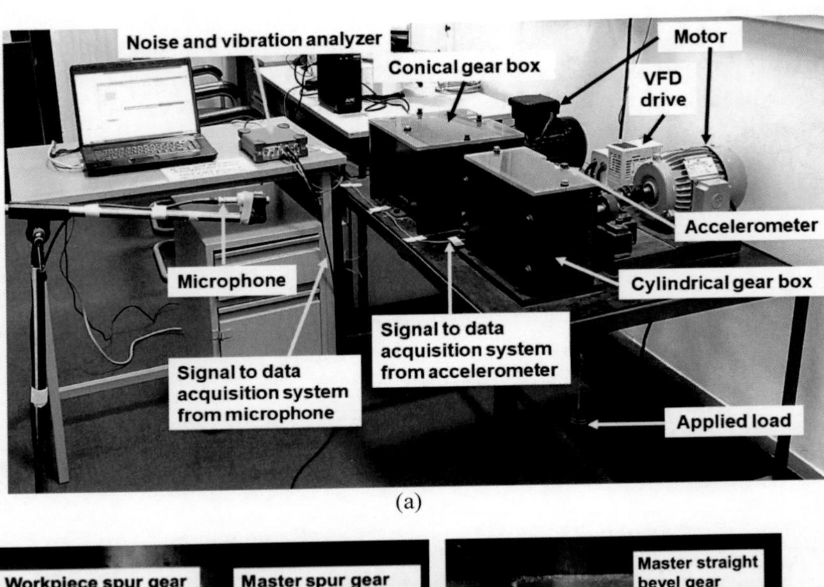

(a)

(b) (c)

Fig. 9.1: Photograph of the (a) equipment and devices used for measurement of noise and vibrations of the unfinished and the Stage-3 experimentation best finished spur and straight bevel gears by the developed test rig, and arrangement of the workpiece and its master gear in it for the (b) spur gear, (c) straight bevel gear (Petare et al., 2018)

9.2 Evaluation of the Functional Performance Parameters of Gears

Functional performance parameters for the unfinished and the stage-3 experimentation best finished spur gears were measured by an in-house developed dual flank roll tester by Kasliwal et al. (2017), which was further modified by providing a stepper motor, Arduino programmed micro-controller, and laser displacement sensor (figure 9.2). It has a provision to mount the workpiece or test gear and its master gear whose theoretical center distance can be adjusted by a lead screw. They are made to mesh with each other, maintaining dual flank contact and are rotated by a stepper motor, whose speed is controlled by an Arduino-based micro-controller. A laser displacement sensor records the variations in the center-to-center distance between the test and the master gears by observing the movement of the plate on which the master gear is mounted. It gives a data file that is used to graphically depict variations of the center-to-center distance with the angle of rotation from which the considered functional performance parameters can be determined.

A computer numerically controlled (CNC) dual flank roll tester (DO-125 KPC from *Gearspect, Pune, India,* please refer to Appendix-B for its details) was used to measure the considered parameters of functional performance for the unfinished and the stage-3 experimentation best finished straight bevel gears. It has a provision to mount the test and master bevel gears at their designed angle of intersection while maintaining dual flank contact. The speed of the rotation of the test gear is controlled by its associated software, which also records the variations in the center-to-center distance between the test and master gears, gives its variation in a plot, and gives computed values of functional performance parameters of conical gears.

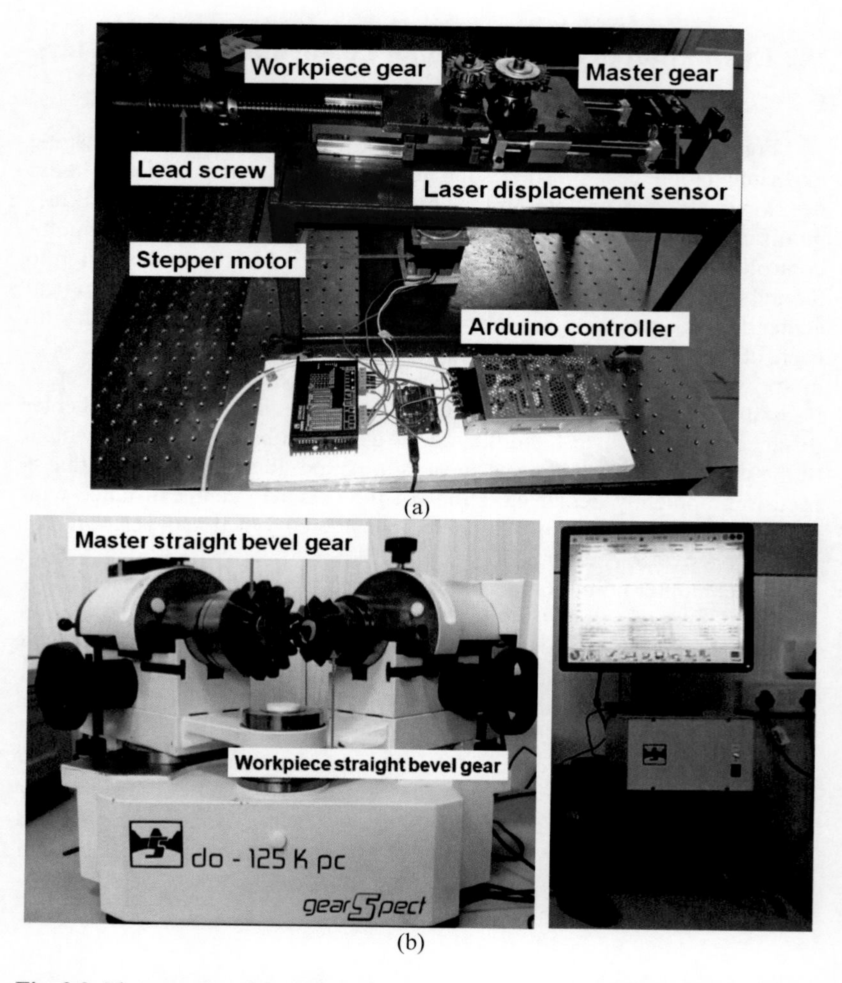

Fig. 9.2: Photographs of the double flank roll tester for (a) spur gears (Kasliwal et al., 2017), and (b) straight bevel gears.

9.3 Results and Analysis of Functional Performance Parameters

The best finished spur and straight bevel gears corresponding to exp. No. 10 of the stage-3 experimentation (refer to chapter 6) was used for dual flank roll testing and noise and vibration testing. Figure 9.3 shows the variations in the center-to-center distance (marked as displacement in μm) with the rotation angle for the unfinished spur (fig. 9.3a) and straight bevel gear (fig. 9.3c), and the corresponding best finished gears (fig. 9.3b, and 9.3d for spur and straight bevel gears, respectively). The values of total composite error, tooth-to-tooth composite error, and radial runout computed from these graphs are shown. It can be observed from these graphs that the AFF process reduced total composite error from 270 to 130 μm (51.9% reduction) for the spur gear and 125 to 84 μm (32.8% reduction) for the straight bevel gear. It reduced tooth-to-tooth composite error from 209 to 125 μm (40.2% reduction) for the spur gear, and from 124 to 84 μm (32.3% reduction) for the straight bevel gear. More importantly, it significantly reduced the dual flank roll testing determined radial runout from 126 to 80 μm (36.5% reduction) for the spur gear, and from 19.7 to 15 μm (23.9% reduction) for the straight bevel gear. This is due to reductions in microgeometry errors for both types of gears, which result in less variation in the center-to-center distance when the workpiece gear is rotated in mesh with its master gear during the dual flank roll testing. It is worth mentioning that the CNC gear metrology-determined percentage reductions in radial runout are 10% and 51.5% for the spur and straight bevel gears, respectively.

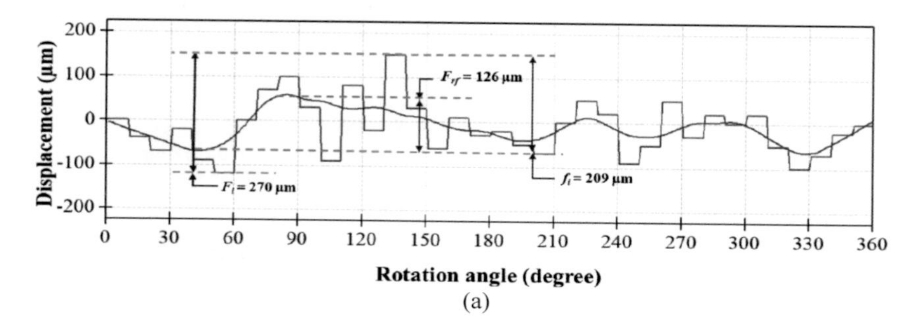

(a)

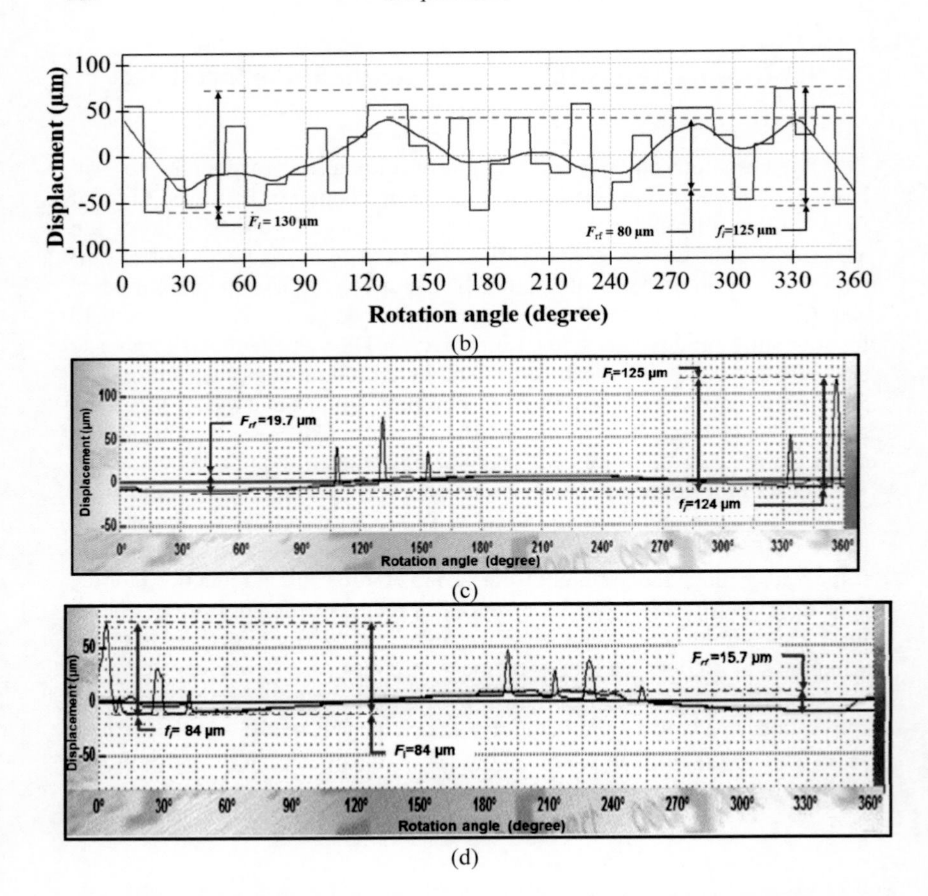

Fig. 9.3: Total composite error and tooth-to-tooth composite error for the (a) unfinished spur gear, (b) Stage-3 experimentation best finished spur gear, (c) unfinished straight bevel gear, and (d) Stage-3 experimentation best finished straight bevel gear (Petare et al., 2018)

9.4 Results and Analysis of Gear Noise and Vibrations

Table 9.2 presents the values of noise and vibrations of the unfinished and best finished spur and straight bevel gears (corresponding to exp. no. 10 of the stage-3 experimentation of chapter 6) for 16 full factorial experiments corresponding to four values of rotational speed and applied to each gear's load. Figures 9.4 and 9.5 graphically depict the changes in noise level and vibration level respectively with the applied load for different rotational speeds for the unfinished and the best finished spur gear (figs. 9.4a and 9.5a) and the unfinished straight bevel gear and the best finished straight bevel gear (figs. 9.4b and 9.5b). The following observations can be made from figs. 9.4 and 9.5, and table 9.2 along with their explanation:

- Stage-3 experimentation best finished gears by the AFF process generate less noise than their corresponding unfinished gears for all the considered values of load and speed. The best finished spur and straight gears showed more noise reduction at 500 rpm and 250 rpm, respectively, than at other speeds. The maximum noise reductions obtained are 5.2 dBA occurring at 5 N load (exp. no. 6 in table 9.2) for spur gear, and 4.6 dBA at 3 N load (exp. no. 1 in table 9.2) for straight bevel gears. These reductions in noise level are due to the reduction in the total profile error of the AFF finished spur gear and the consequent DIN quality improvement (DIN 12 to DIN 11 as mentioned in table 6.7; please refer to chapter 6) and surface roughness for both the spur and straight bevel gears by the AFF process.
- Vibration levels of the stage-3 experimentation best finished spur and straight bevel gears are found to be lower than their corresponding unfinished gears at all the considered values of load and speed (figs. 9.5a and 9.5b). The amount of vibration reduction for the best finished gears increases with speed. Vibrations of the unfinished and the best finished spur gears were much higher at 1000 rpm, for all the values of the load than corresponding values at lower speeds. The obtained values of the maximum vibration reductions are 5.3 m/s^2 and 4 m/s^2 for the spur and straight bevel gears, respectively, occurring at 1000 rpm speed and 9 N load (exp. no. 16 in table 9.2). These reductions in vibrations are due to the reductions in total pitch error (DIN quality improved from 10 to 9 for both the AFF best finished spur and straight bevel gears, as mentioned in table 6.7), radial runout (DIN quality improved from 11 to 9 for the AFF best finished straight bevel gear and reduced from 54.7 to 49.2 μm for the AFF best finished spur gear, as mentioned in table 6.7, and errors in functional performance parameters after finishing by the AFF process.

Table 9.2: Values of noise and vibrations of the unfinished and the stage-3 experiment's best finished spur and straight bevel gears for different combinations of rotational speed and applied load

Exp. No.	*Rotational speed (rpm)*	*Applied load (N)*	*Spur gear*						*Straight bevel gear*					
			Noise level (dBA)			*Vibration (m/s^2)*			*Noise level (dBA)*			*Vibration level (m/s^2)*		
			Unfinished	*Finished*	*ΔN*	*Unfinished*	*Finished*	*ΔV*	*Unfinished*	*Finished*	*ΔN*	*Unfinished*	*Finished*	*ΔV*
1	250	3	82.3	81.1	1.2	2.5	2.6	-0.1	**79.4**	**74.8**	**4.6**	2.6	1.9	0.7
2		5	82.8	78.5	4.3	2.7	2.2	0.4	76.5	75.5	1.0	2.2	2.0	0.2
3		7	83.3	79.3	4.0	2.8	2.5	0.3	78.7	76.9	1.8	2.9	2.2	0.7
4		9	83.6	79.2	4.4	2.9	2.5	0.4	76.9	76.7	0.2	2.4	2.1	0.2
5	500	3	92.2	89.1	3.1	9.2	8.2	1.0	84.3	82.6	1.7	4.6	4.2	0.5
6		**5**	**92.0**	**86.8**	**5.2**	9.2	6.3	2.9	85.4	84.2	1.2	5.3	5.4	-0.1
7		7	91.9	87.5	4.4	9.4	6.4	3.0	85.3	83.6	1.7	5.9	4.9	1.0
8		9	91.8	88.3	3.5	9.3	7.0	2.3	85.2	86.3	-1.1	6.9	5.9	1.0
9	750	3	93.9	93.3	0.6	17.4	13.9	3.5	90.2	89.5	0.7	10.9	8.7	2.2
10		5	93.4	93.6	-0.2	16.2	14.3	2.0	90.9	89.7	1.2	10.7	10.5	0.2
11		7	94.2	93.8	0.4	17.1	15.2	1.9	89.9	89.7	0.2	10.6	9.8	0.8
12		9	95.4	94.0	1.4	17.5	16.4	1.0	90.8	89.6	1.2	12.5	9.8	2.7
13	1000	3	97.3	95.9	1.4	24.9	20.5	4.4	93.7	93.2	0.5	15.4	14.0	1.5
14		5	97.8	95.8	2.0	24.9	20.7	4.2	94.1	93.6	0.5	16.3	15.0	1.3
15		7	97.7	96.1	1.6	25.1	20.8	4.4	94.0	93.8	0.2	16.2	14.1	2.1
16		9	97.6	96.2	1.4	**26.0**	**20.6**	**5.3**	96.1	93.7	2.4	**18.2**	**14.2**	**4.0**

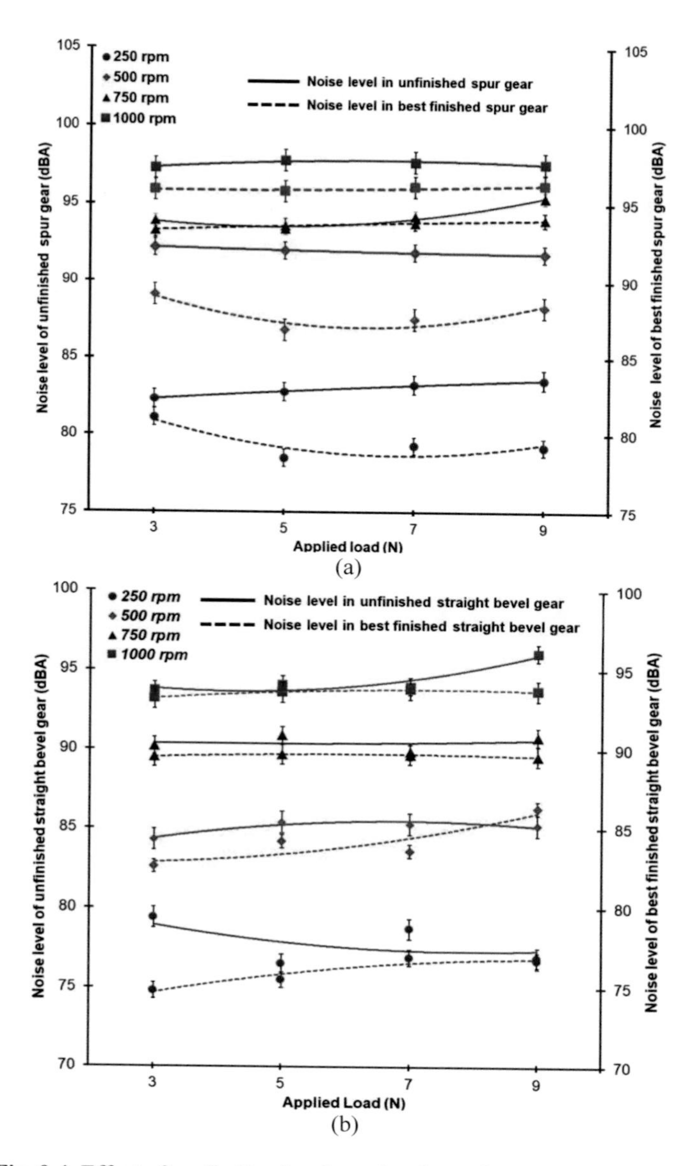

Fig. 9.4: Effect of applied load and rotational speed on noise level for the unfinished and the stage-3 experimentation best finished (a) spur gear, and (d) straight bevel gear

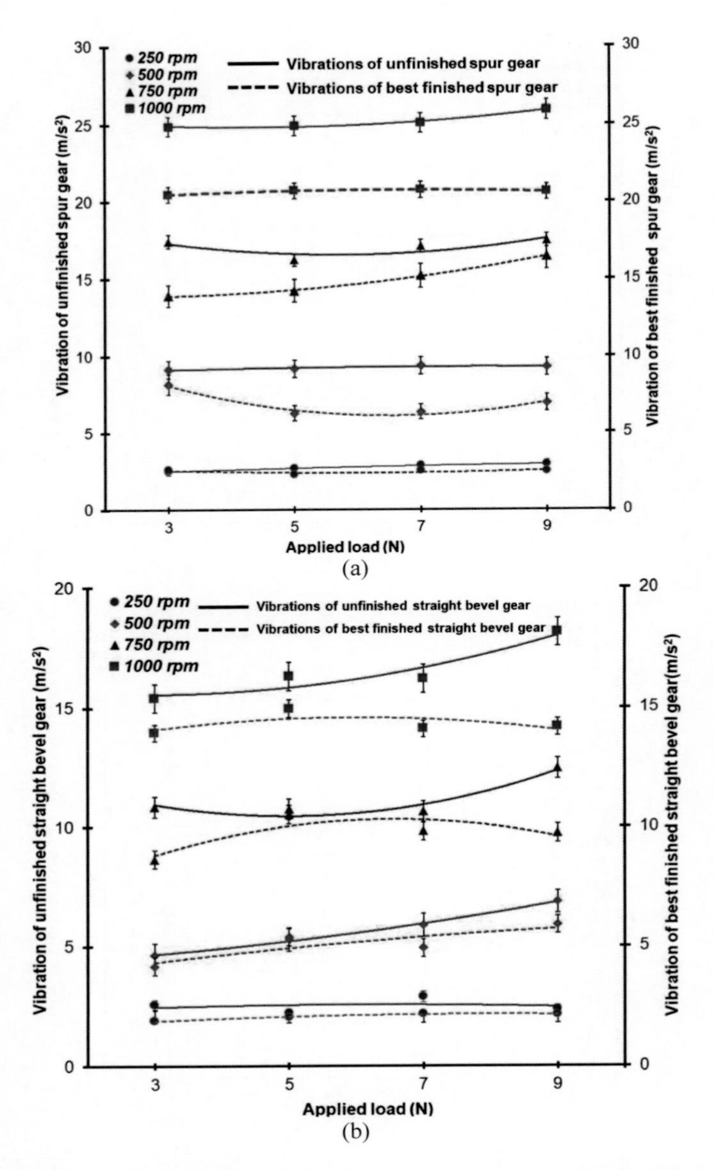

Fig. 9.5: Effect of applied load and rotational speed on the vibration of the unfinished and the stage-3 experimentation best finished (a) spur gear, and (b) straight bevel gear

9.5 Concluding Remarks

The following conclusions can be drawn based on the performance characteristics study of the spur and straight bevel gears after their finishing by the AFF processes:

- Finishing of both spur and straight bevel gears by the AFF process considerably reduces the (a) total profile error and total lead error of the spur gear, which reduce its noise and enhances the load carrying capacity, respectively; (b) total pitch error for both types of gears, enhancing their motion transfer characteristics; (c) radial runout, which reduces their vibration level and non-uniform motion, and (d) max. and avg. surface roughness, which results in their improved service life and operating performance and also reduces their noise and vibrations.
- The AFF process also reduces the functional performance parameters for both types of gears. The percentage reduction in radial runout determined by the dual flank roll testing for the spur gear is more than that determined during its measurement of microgeometry errors, whereas the values of radial runout determined by the dual flank roll testing for the straight bevel gear are significantly less than that determined during its measurement of microgeometry errors. This is very significant because roll testing simulates the actual working conditions of a test gear by meshing it with its master gear, unlike the microgeometry error measurement in which measurements are done only on the test gear without meshing it with any other gear. This will result in the uniform transmission of motion or power and fewer vibrations generated by the AFF-finished gears.
- Gear finishing by the AFF process reduced the noise and vibrations of both types of gears due to enhancing their surface characteristics and reducing their microgeometry errors. The reduction in surface roughness improves the contact area of a gear with its meshing gear, giving a more uniform distribution of the forces and resulting in fewer vibrations.

References

[1] Kashyap, R, N. K. Jain and A. Parey. 2017. *Analysis and Comparison of Noise and Vibrations of Straight Bevel Gears Finished by Advanced Finishing Processes*, M.Tech Thesis, Discipline of Mechanical Engineering, IIT Indore. http://dspace.iiti.ac.in:8080/ jspui/handle/123456789/481

[2] Kasliwal, R, N. K. Jain and A. Parey. 2017. *Design and Development of Double Flank Roll Tester for Cylindrical Gears,* M.Tech. Thesis, Discipline of Mechanical Engineering, IIT Indore. http://dspace.iiti.ac.in:8080/jspui/handle/123456789/747

[3] Kumar, G, Jain, N. K., Parey, A. 2017. *Analysis and Comparison of Noise and Vibrations of Spur Gears Finished by Advanced Finishing Processes*, M.Tech. Thesis, Discipline of Mechanical Engineering, IIT Indore. http://dspace.iiti.ac.in:8080/jspui/handle/123456789/523

CHAPTER TEN

CONCLUSIONS AND FUTURE RESEARCH AVENUES

This chapter summarizes the significant outcomes and conclusions from the work presented in this book along with future research avenues.

10.1 Significant Outcomes

The following are some significant outcomes of the work on gear quality enhancement by the AFF process:

- Successful development of the AFF machine for finishing spur and straight bevel gears, having the facility for automatic operation and a stroke counter.
- Development of the fixtures for holding and supporting gears of different types to enable their finishing by the AFF process.
- Extensive investigation to improve the microgeometry and surface finish of spur and straight bevel gears by the AFF process, which enhanced their surface morphology, microhardness, wear characteristic, functional performance aspects, and noise and vibration characteristics.
- Sophistication of the manual dual flank roll tester for cylindrical gears using a stepper motor with an interface with an Arduino-based microcontroller to control the movement of the mating gears.
- Use of laser texturing to improve the finishing performance and productivity of the AFF process.
- Study on the reduction of noise and vibrations and functional performance parameters of spur and straight bevel gears after their finishing by the AFF process.

10.2 Conclusions

The following are the main conclusions drawn from the present work:

- **Stage-1 experimentation:** Limiting value of finishing time at 30 minutes and extrusion pressure as 7.5 MPa identified during this stage of experimentation because the surface roughness values of both spur and straight bevel gears were found to decrease beyond these values. The identified feasible range of finishing time and extrusion pressure were used in further experiments.
- **Stage-2 experimentation:** (i) Both finishing time and concentration of silicone oil (used as a measure of the viscosity of the AFF medium) were found to have a significant impact on the simultaneous improvement in surface finish, wear characteristics, and microgeometry of spur and straight bevel gears during their finishing by the AFF process; (ii) Reductions in surface roughness parameters and microgeometry errors continuously increase with the increase in viscosity of the AFF medium (i.e. decrease in vol.% of the blending oil) and finishing time, and attain their maximum values at a finishing time of 25 minutes for spur gears and 30 minutes for straight bevel gears, and 10% vol. concentration of silicone oil (i.e. 135 kPa.s as the viscosity of the AFF medium); (iii) Blending oil plays an important role in the proper mixing of the abrasive particles with the putty and to hold them together against the extrusion pressure in the AFF process. Reducing its vol. concentration below 10% to further increase viscosity of the AFF medium can choke the AFF medium-containing cylinders and the restriction between the workpiece gear and its fixture. This results in insufficient slug formation due to poor bonding between the putty and abrasive particles. Therefore, 10% vol. concentration of silicone oil for both the spur and straight bevel gears and a finishing time of 25 minutes for spur gears and 30 minutes for straight bevel gears were identified as optimum values for further experiments.
- **Stage-3 experimentation:** (i) Increases in the extrusion pressure and size of the abrasive particles continuously increase reductions in microgeometry errors and surface roughness parameters of spur and straight bevel gears, and with attainment of their maximum values of extrusion pressure (i.e. 7 MPa) and size of the abrasive particles (i.e. 120 mesh); (ii) An optimum value of 30% exists for vol. concentration of abrasive particles in the AFF medium

because reductions in microgeometry errors and surface roughness parameters of spur and straight bevel gears initially increase up to this value and then start decreasing, (iii) Therefore, 7 MPa extrusion pressure, 120 mesh size (127 μm avg. diameter) of abrasive particles, and 30% vol. concentration of abrasive particles is identified as the optimum parametric combination by the RSM for further study.

- **Optimization by the DFA and experimental validation:** (i) There is very good agreement between the optimum values of the considered responses obtained by the validation experiment with those predicted by the DFA and identified by the RSM. Prediction error for the DFA lies in the range from -9.7 to 10.3 for spur gears and from 1.7 to 8.9 for straight bevel gears, i.e. prediction error is higher for spur gears than straight bevel gears; (ii) DFA overpredicts values of % reductions in total profile error, total lead error and total pitch error of AFF-finished spur gears, and % reductions in total pitch error, radial runout and avg. surface roughness of AFF-finished straight bevel gears compared to those predicted by the validation experiments and RSM. The RSM overpredicts the % reductions in the radial runout of the AFF-finished spur gear and in the max. surface roughness of the AFF-finished straight bevel gear; (iii) DFA underpredicts the % reductions in the max. and avg. surface roughness of the AFF-finished spur gear but overpredicts the % reduction in microgeometry errors in spur gears and all the considered responses for straight bevel gears; (iv) The AFF of spur and straight bevel gears using the optimized parameters makes their bearing area more uniform compared to their unfinished state due to reductions in their max. and avg. surface roughness. Improvement in the bearing area will improve their wear characteristics, which will result in a better tribological performance during their service life.
- **Study of surface integrity aspects of the best-finished gears:** (i) Surface morphology study of the AFF-finished spur and straight bevel gears revealed that their flank surfaces are free from gear cutter marks, burrs, nicks, microcracks, microchips, and micro-pits due to the restricted and concentrated movement of the AFF medium to the desired finishing area. It also revealed that finishing action by AFF is due to an abrading action followed by micro-cutting with plowing on the gear teeth flank surfaces, (ii) The AFF process significantly reduced the maximum value of sliding friction force, coefficient of sliding friction, specific wear rate,

wear rate, and sliding wear volume of spur and straight bevel gears. A surface morphology study of the worn flank surfaces of the AFF-finished spur and straight bevel gears revealed material displacement and flaking, indicating their wear by scuffing mode, (iii) The AFF process increased the microhardness of spur and straight bevel gears by a significant amount due to the use of higher extrusion pressure, which exerts axial and radial forces on the abrasive particles and induces compressive residual stresses, thus improving the fatigue life of their flank surfaces and simultaneously imparting a very fine finish to them by abrasion; (iv) Reduction of surface roughness of gear flank surface improved the bearing area, which enhances the wear characteristics of the AFF-finished spur and straight bevel gears.

- **Laser texturing of gears:** (i) Laser power, focal length, and number of passes significantly influence the homothetic texture of the flank surfaces of a workpiece gear in a direction perpendicular to the lay pattern generated by the gear manufacturing process. A combination of 20 W fiber laser, 285 mm focal length, and five passes was found to be the optimum values in the present study; (ii) The AFF of *laser-textured* spur and straight bevel gears gives higher values of vol. MRR and a higher percentage of reductions in their microgeometry errors and surface roughness parameters than the AFF of *untextured* spur and straight bevel gears for all the values of finishing time. This is due to the formation of a mesh-like structure between the homothetic laser textures and lay pattern of a gear cutter, which deflects the flow of the abrasive particles during finishing and provides them with a random motion with more active abrasive particles. It enables them to cover a longer flow path over the flank surfaces, giving more uniform abrasion, more MRR, and more reductions in the considered responses in less finishing time; (iii) The surface morphology study of flank surfaces of the AFF-finished laser-textured gears reveals smoother surfaces than the flank surfaces of the AFF- finished untextured gears; (iv) The AFF process yields more microhardness in the AFF-finished laser-textured gears than that of the AFF-finished untextured gears due to the continuous impact of the abrasive particles caused by the extrusion pressure and the formation of subgrain boundaries by laser texturing; (v) The AFF-finished laser-textured gears offer a higher wear resistance compared to the AFF-finished untextured gears due to their smoother surfaces, with less variation in the roughness profile produced by the AFF process and distortion of lattice structure produced by laser texturing, which forms compressive residual stresses; (vi) Laser texturing is an effective process to enhance the productivity of the AFF process in terms of improvements in microgeometry, surface finish, surface morphology, wear resistance, microhardness, and MRR.

- **Performance characteristics of gears:** (i) Finishing of both spur and straight bevel gears by the AFF process considerably reduces the (a) total profile error and total lead error of spur gears, which reduce their noise and enhance their load-carrying capacity, respectively; (b) total pitch error for both types of gears, enhancing their motion transfer characteristics and transmission accuracy; (c) radial runout, which reduces their vibration level and non-uniform motion; and (d) max. and avg. surface roughness of gears, improving their service life and operating performance and also reducing their noise and vibration; (ii) Gear finishing by the AFF process resulted in a maximum reduction noise level for the AFF-finished spur gear by 5.2 dBA and the AFF-finished straight bevel gear by 4.6 dBA. Similarly, the maximum vibration level for the AFF- finished spur gear was reduced by 5.3 m/s^2 and for the AFF-finished straight bevel gear by 4 m/s^2. This is due to the enhancement in their surface characteristic and the reduction in their microgeometry errors. Reduction in surface roughness improves the contact area of a test gear, with its meshing gear giving more uniform distribution of the forces and resulting in fewer vibrations; (iii) The AFF of both types of gears also reduced their functional performance parameters. Percentage reduction in radial runout determined by the dual flank roll testing for spur gears is more than that determined during its measurement of microgeometry errors, whereas the values of radial runout determined by the dual flank roll testing for straight bevel gears are significantly less than that determined during its measurement of microgeometry errors. This is very significant because roll testing simulates the actual working conditions for a test gear by meshing it with a master gear, unlike the microgeometry error measurement in which measurements are done only on the test gear without meshing it with any other gear. This will result in the uniform transmission of motion or power and fewer vibrations by the AFF-finished gears.
- This study proves that AFF is an economical, productive, flexible, and sustainable alternative advanced process for gear finishing that can simultaneously improve microgeometry, surface finish,

surface morphology, wear characteristics, and microhardness, and reduce noise and vibration and errors in functional performance parameters of all types of gears.

10.3 Future Research Avenues

The present work is the first attempt to establish the AFF process as an alternative gear finishing process to improve gear microgeometry, surface roughness, surface morphology, microhardness, wear characteristics, and functional performance aspects of the spur and straight bevel gears. Therefore, there is a lot of scope for future research avenues:

- Explore the AFF process for imparting different types of modifications to gear teeth such as tip and root relieving, end relieving, flank crowning, profile crowning, modification of pressure, and helix angle.
- Investigation of the simultaneous improvement of the microgeometry and surface finish of the spiral bevel gear, rack and pinion gear, worm and worm wheel, non-circular gears, and their sound intensity analysis.
- Explore other hybrid and derived variants of the AFF process for finishing gears and develop flexible fixtures to hold and finish different types and sizes of gears.
- Explore the micro-plasma transferred arc (μ-PTA) process or other processes for texturing gears to improve the productivity of the AFF process.

APPENDIX-A

CHEMICAL COMPOSITION OF THE GEAR MATERIAL

Certificate

CHOKSI LABORATORIES LIMITED
(ANALYTICAL TESTING & CALIBRATION LABORATORY GROUP)

6/3, MANORAMAGANJ, INDORE 452001 MADHYA PRADESH
Phone : +91 731 4243888 (30 LINES)
Fax : +91 731 2490593
Email : indore@choksilab.com
Website : www.choksilab.com

BOOKING NO. : 10349
CERTIFICATE NO. : 10349/2017-2018

1. NAME OF MANUFACTURER/PARTY : SUJEET KUMAR CHAUBEY
RESEARCH SCHALAR
IIT,INDORE
INDORE MADHYA PRADESH

1.MFG. LIC. NO.	: NM	5. REFERENCE NO. 1	: NM
2. REFERENCE NO. OF LETTER	: NM	6. REFERENCE NO. 2	: NM
3. DATE	: NM	7. REFERENCE NO. 3	: NM
4. DATE OF RECEIPT	: 14/10/2017	8. NAME OF SAMPLE	: METAL PIECE (20 MnCr5)

9. DETAILS OF RAW MATERIAL / FINAL PRODUCTS (In Bulk/Finished Pack)

(A) BATCH NO.	: 2	(G) SAMPLE QUANTITY	: 1 NO
(B) BATCH SIZE	: NM	(H) SEALED	: UNSEALED
(C) A.R. NO.	: NM	(I) PACKING	: OPEN
(D) DATE OF MFG.	: NM	(J) STARTING DATE	: 14/10/2017
(E) DATE OF EXPIRY	: NM	(K) ENDING DATE :	: 14/10/2017
(F) MFG NAME	: NM	(L) PAGE NO.	: Page 1 of 1

AS PER CUSTOMER SPECIFICATION

SR	TEST NAME	UNIT	RESULT	SPECIFICATIONS	METHOD OF TEST
(A)	Chemical Composition : --	---			
1	Carbon as C	%	0.189	0.17 to 0.22	CLL-IDR-STP-BM-004
2	Silicon as Si	%	0.249	Max. 0.40	(Spectro Analysis)
3	Manganese as Mn	%	1.19	1.10 to 1.40	
4	Phosphorus as P	%	0.019	Max. 0.035	
5	Sulphur as S	%	0.017	Max. 0.035	
6	Chromium as Cr	%	1.10	1.00 to 1.30	

Remarks :
1. Certificate Issue Date: 14/10/2017

For Choksi Laboratories Ltd.
Authorised Signatory

Sachin Upadhyay
Sr. Scientist
Building Materials
AUTHORISED SIGNATORY
QAIS : 57/01

Note :
Please see overleaf for terms & conditions.

APPENDIX-B

DETAILS OF THE MEASURING INSTRUMENTS USED

- **Scanning Electron Microscope (Sophisticated Instrument Centre, IIT Indore)**

Make	Carl Zeiss NTS GmbH, Germany
Model	SUPRA 55
Resolution	1.0 nm @ 15 kV 1.7 nm @ 1 kV 4.0 nm @ 0.1 kV
Acceleration Voltage	0.1 - 30 kV
Magnification	12x - 900,000 x
Stages	5-Axes Motorized Eccentric Specimen Stage X = 130 mm, Y = 130mm, Z = 50mm, T = -3 - +70° R = 360° (continuous)
Standard Detectors	High-efficiency in-lens detector Everhart-Thornley secondary electron detector

- **CNC Gear Metrology Machine (Gear Research Laboratory, IIT Indore)**

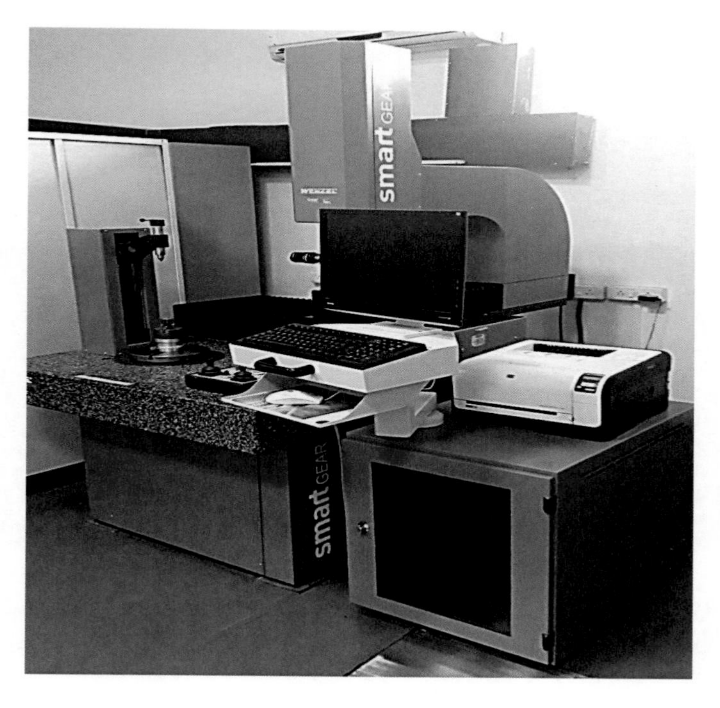

Model	SmartGear 500
Make	WenzelTec Germany
Diameter of work-piece	Minimum/maximum 5-270 mm
Helix angle	<90°
Internal Gear Diameter	>12 mm
Module range	Minimum/maximum 0.4-15 mm
ISO 10360-2 accuracy for 3D measurement from	MPEe = 4.5µm MPEthp = 5.0 µm
Measurable face width	300 mm
Temperature range	20 °C +5k, -3k
Transverse Distance	500 mm (X) 450 mm (Y) 400 mm (Z)
Table diameter	L200 mm

- **3D Surface Roughness cum Contour Tracer (Gear Research Laboratory, IIT Indore)**

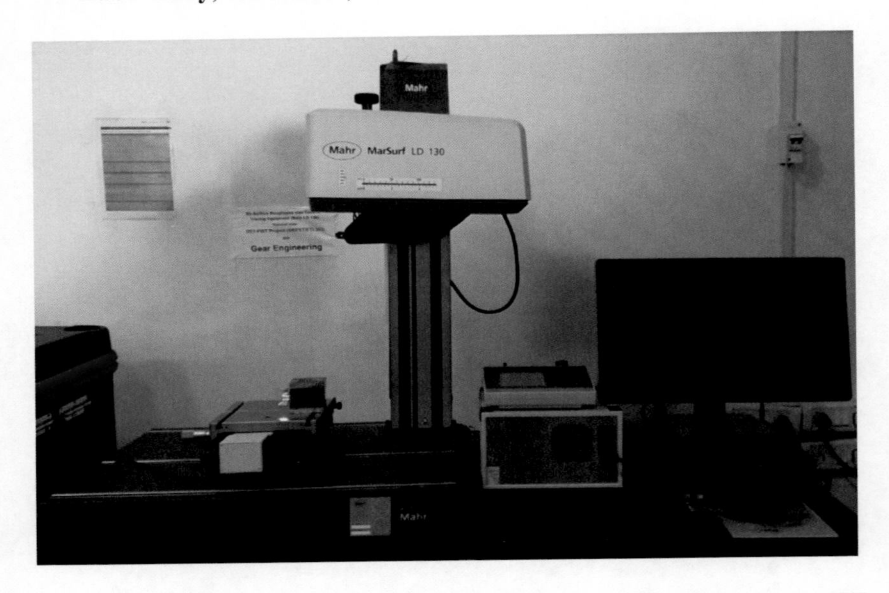

Make	Mahr GmbH
Model	MarSurf LD 130
Resolution	0.8 nm
Start of traversing length (in X)	0.1 mm
End of traversing length (in X)	130 mm
Positioning speed	0.02 mm/s to 200 mm/s
Measuring speed	0.02 mm/s to 10 mm/s for roughness measurements 0.1 mm/s to 0.5 mm/s is recommended
Measuring range (mm)	13 mm (100 mm probe arm) 26 mm (200 mm probe arm)
Traversing lengths	0.1 mm - 130 mm
Measuring force (N)	0.5 mN to 30 mN, software-adjustable

- **Microhardness Tester (Solid Mechanics Laboratory, IIT Indore)**

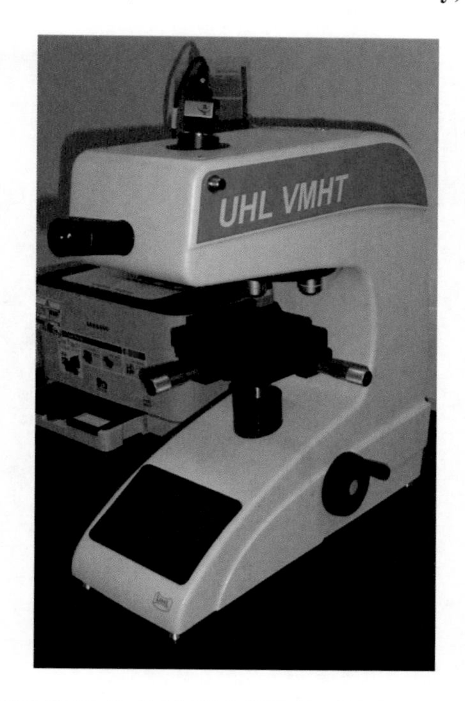

Make	Walter Uhl techn. Mikroskopie GmbH
Model	UHL-VMH-002
Load range	50 grams – 1000 grams
Dwell time	5-99 seconds
Stage size	135 x 135 mm
Type of indenter:	Diamond square base hexagonal pyramid

- **Double Flank Roll Tester for Bevel Gear (Gear Research Laboratory, IIT Indore)**

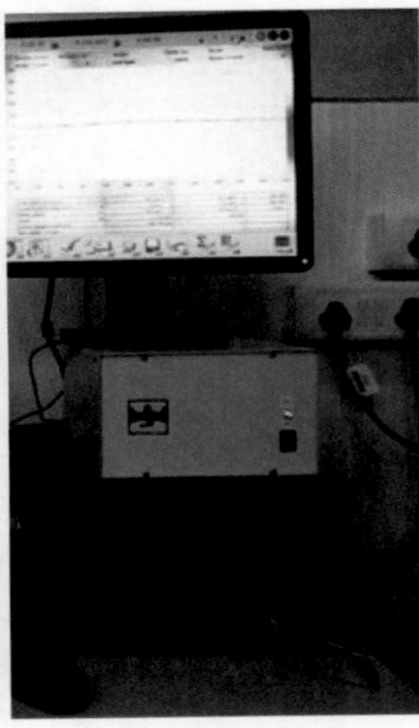

Make	Gearspect India Pvt. Ltd. Pune, India
Model	DO 125 K PC
Min. dia. of measured pinion	10 mm
Max. dia. of measured wheel	125 mm
Max. assembly distance	100 mm
Angle of conical gearing axes	70-120 °

- **Noise and Vibration Analyzer (Gear Research Laboratory, IIT Indore)**

Make	OROS, Grenoble, France
Model	OR 35
No. of channel	4
Software	NV Gate 9.0, 3-series

Make	Dytran Instruments Inc.Chatsworth ,USA
Model	3214A2
Sensitivity	97.86 mV/g

Make	Microtech Gefell GmbH
Model	MV210
Sensitivity	42.9 mv/Pa

- **Rheometer (Sophisticated Instrument Centre, IIT Indore)**

Make	Anton Paar GmbH
Model	MCR 301
Maximum torque	4 mNm
Min. torque, rotation	1 nNm
Min. torque, oscillation	0.5 nNm
Angular deflection	0.05 to ∞ μrad
Angular velocity	10^{-9} to 314 rad/s
Max. speed	3000 rpm

- **Linear Reciprocating Tribometer (Tribology and Metallography Laboratory, IIT Indore)**

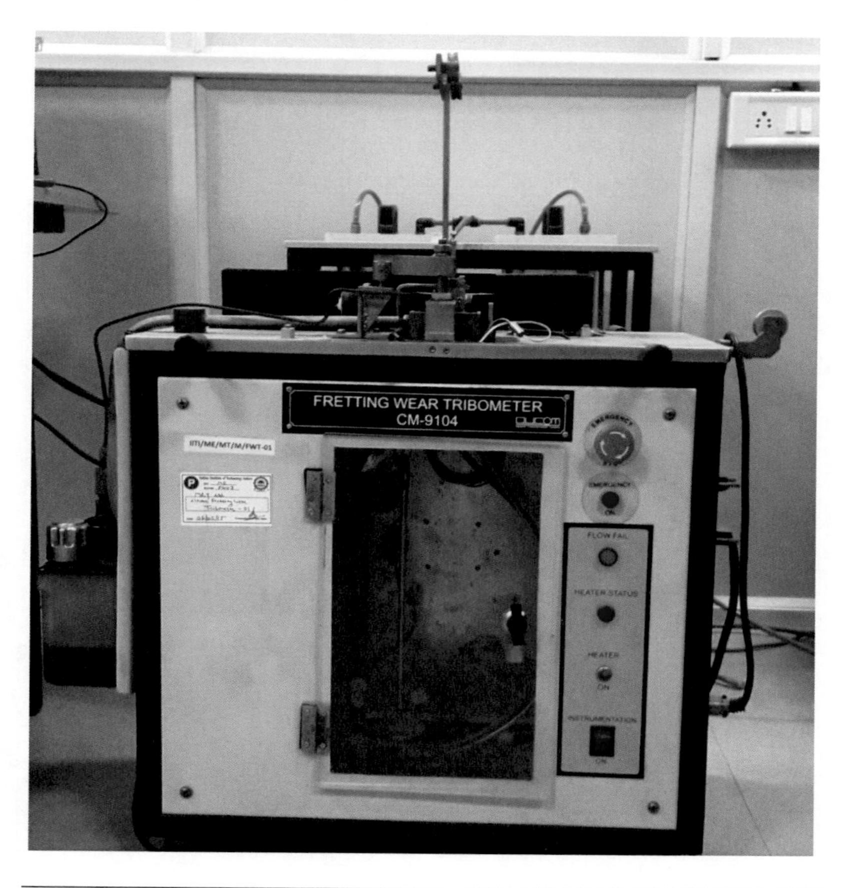

Make	Ducom Instruments, Bangalore, India
Model	CM-9104
Load range	Up to 200 N
Frequency	4 to 40 Hz
Stroke range	1 mm to 30 mm
Compound wear measurement	0 to 1200 μm

APPENDIX –C

EVALUATION GRAPHS OF MICROGEOMETRY ERRORS

(i) Microgeometry graph of Stage 2 experiments

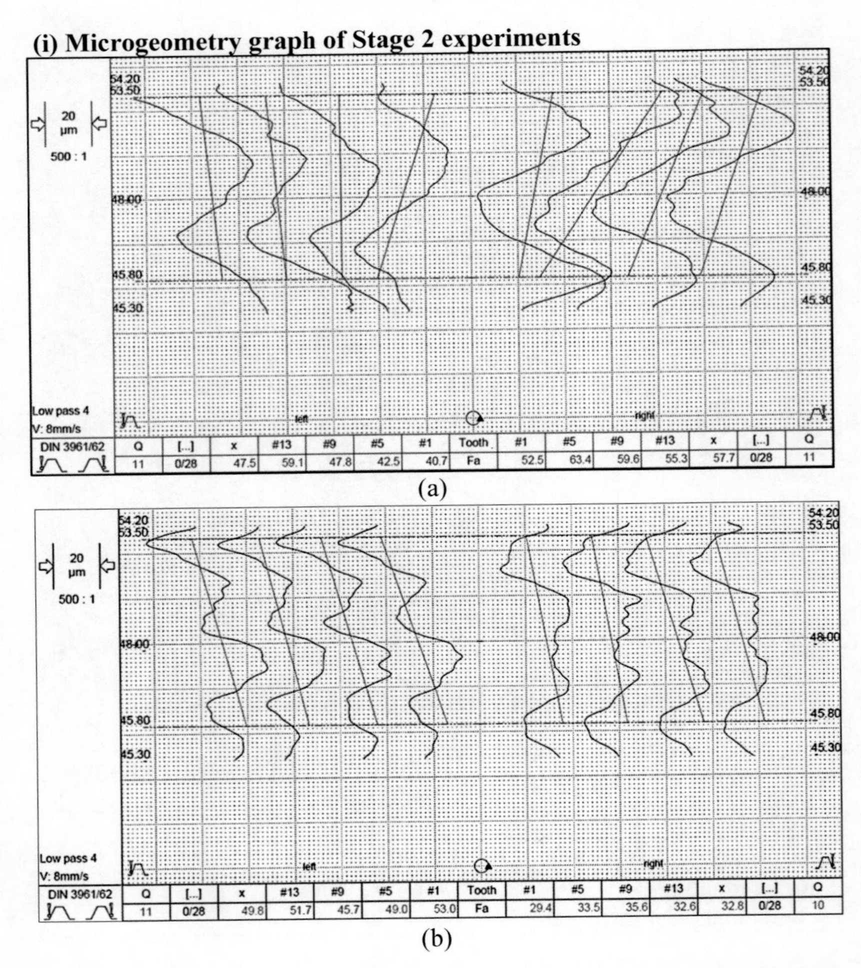

Total profile error 'F_a' of the (a) unfinished spur gear, and (b) Stage-2 best finished spur gear

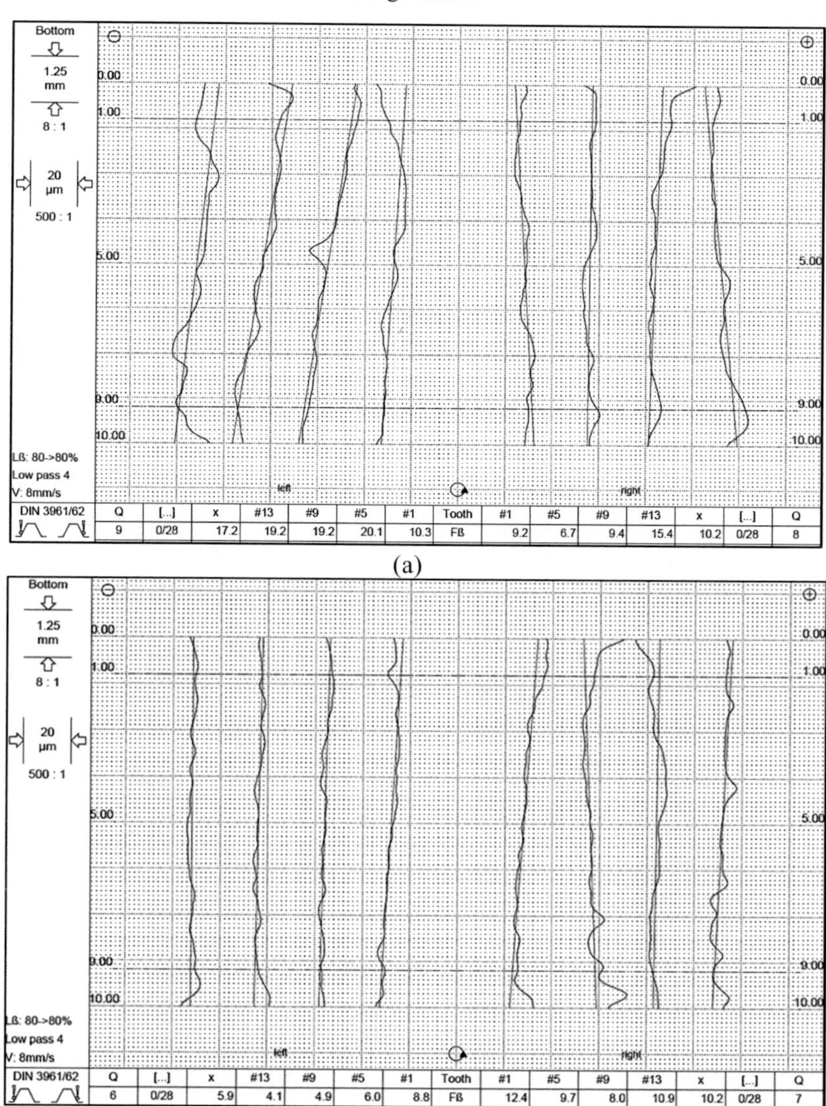

Total lead error 'F_{β}' of the (a) unfinished spur gear, and (b) Stage-2 best finished spur gear

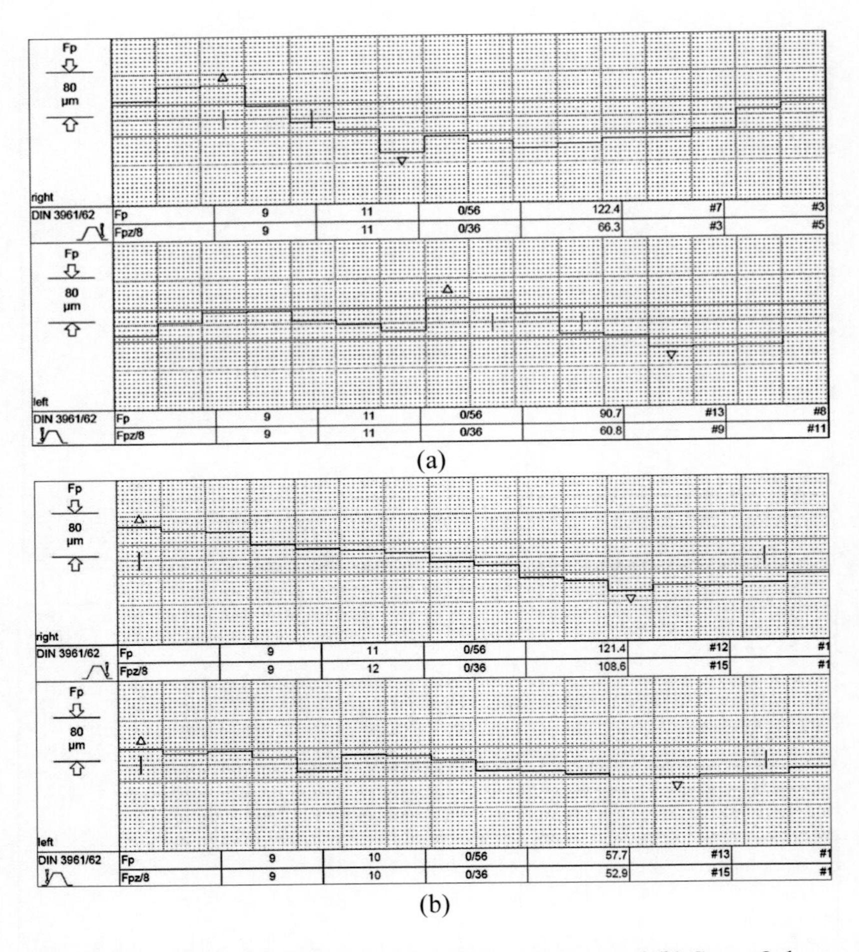

Total pitch error 'F_p' of the (a) unfinished spur gear, and (b) Stage-2 best finished spur gear

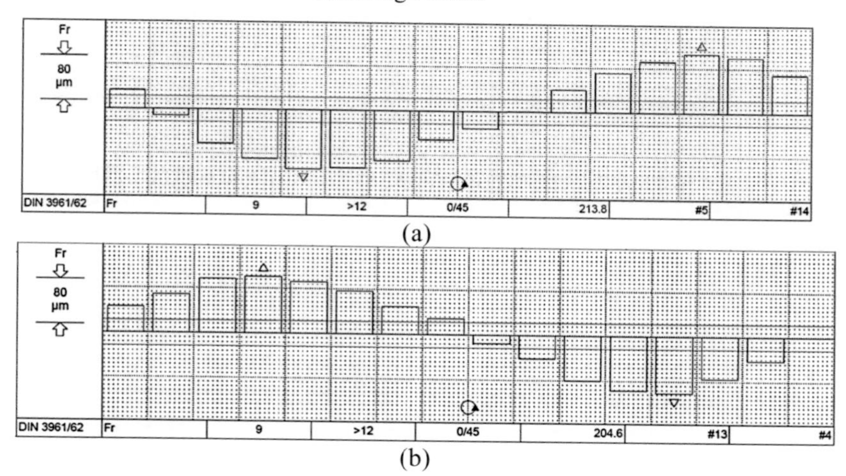

Radial runout 'F_r' of the (a) unfinished spur gear, and (b) Stage-2 best finished spur gear

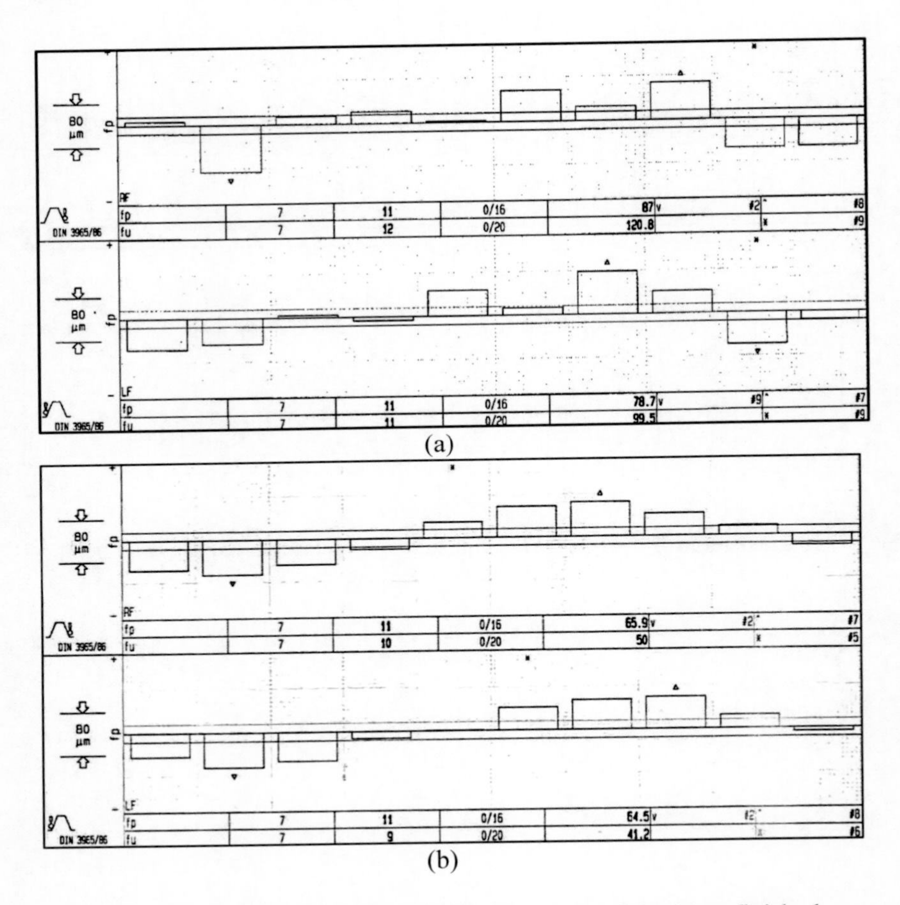

(a)

(b)

Single pitch error 'f_p' and adjacent pitch error 'f_u' of the (a) unfinished straight bevel gear, and (b) Stage-2 best finished straight bevel gear

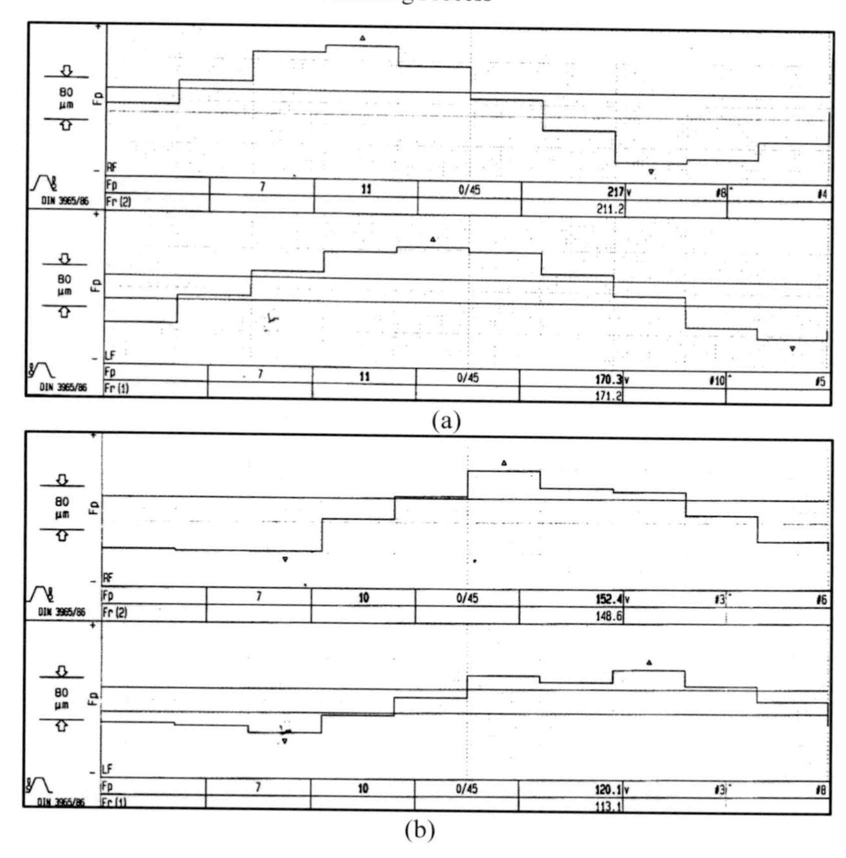

Total pitch error 'F_p' of the (a) unfinished straight bevel gear, and (b) Stage-2 best finished straight bevel gear

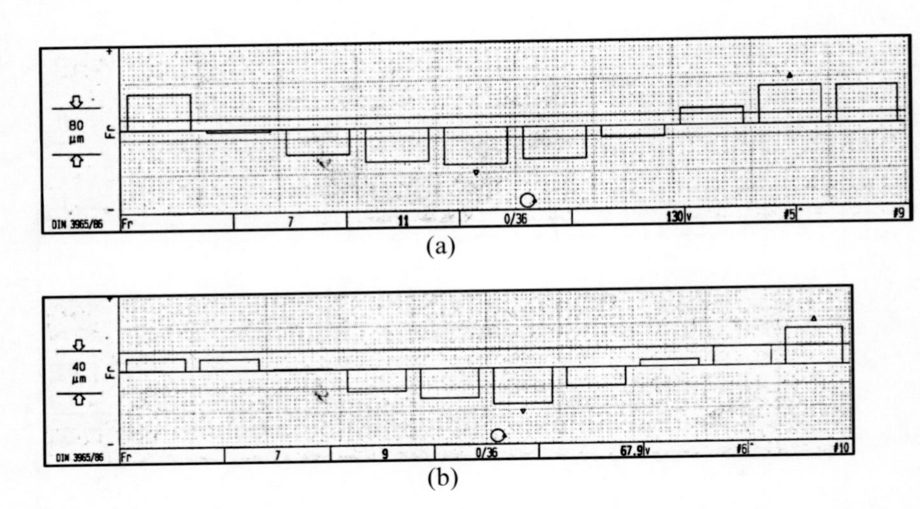

(a)

(b)

Radial runout 'F_r' of the (a) unfinished straight bevel gear, and (b) Stage-2 best finished straight bevel gear

(ii) Microgeometry graph of Stage 3 experiments

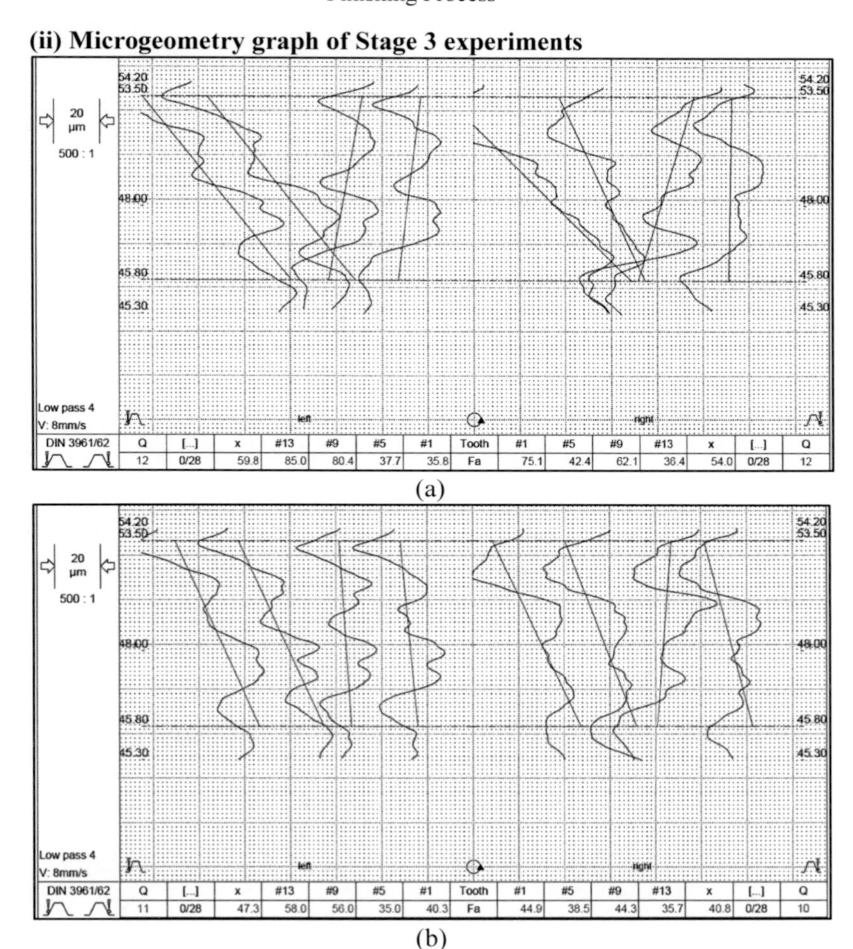

(a)

(b)

Total profile error *'F_a'* of the (a) unfinished spur gear, and (b) Stage-3 best finished spur gear

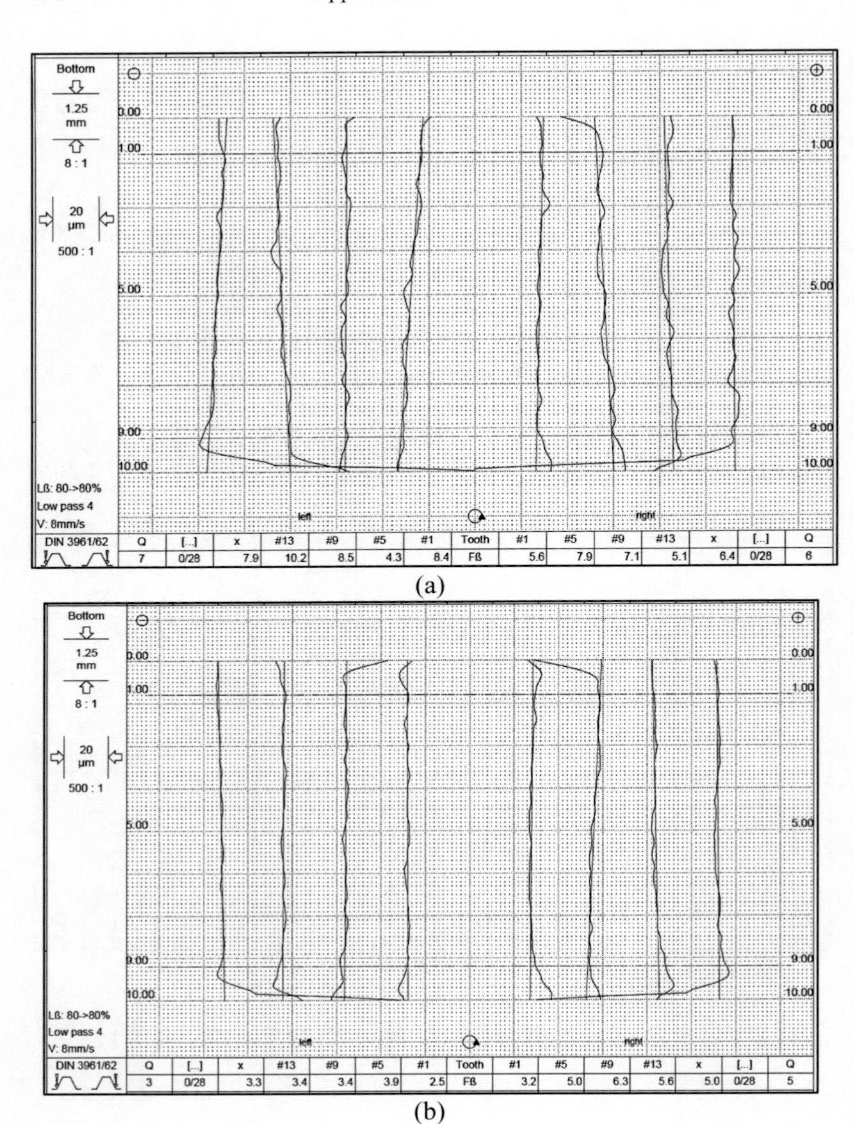

Total lead error 'F_β' of the (a) unfinished spur gear, and (b) Stage-3 best finished spur gear

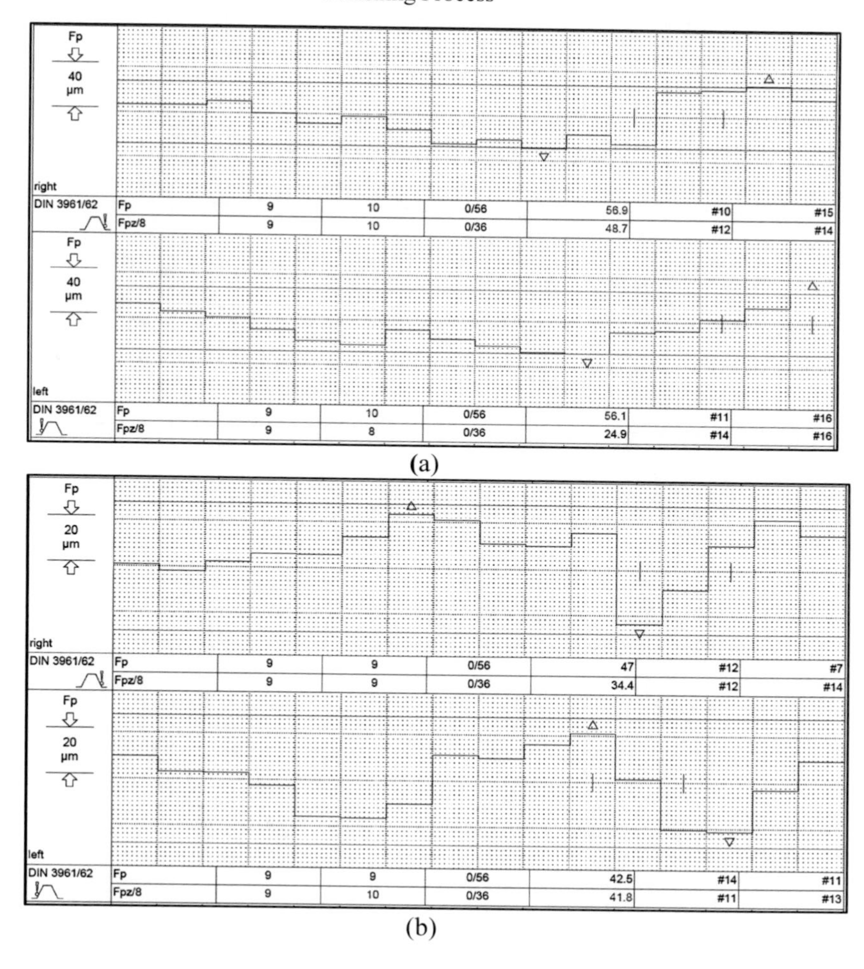

Total pitch error 'F_p' of the (a) unfinished spur gear, and (b) Stage-3 best finished spur gear

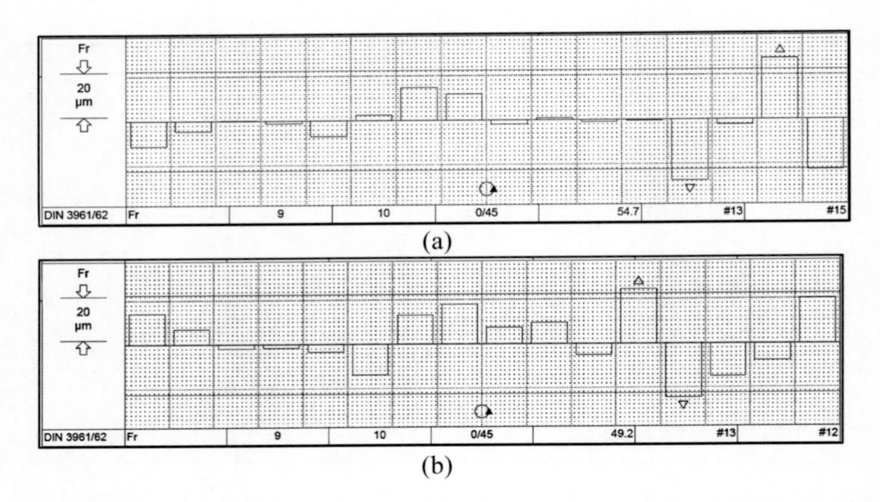

Radial runout 'F_r' of the (a) unfinished spur gear, and (b) Stage-3 best finished spur gear

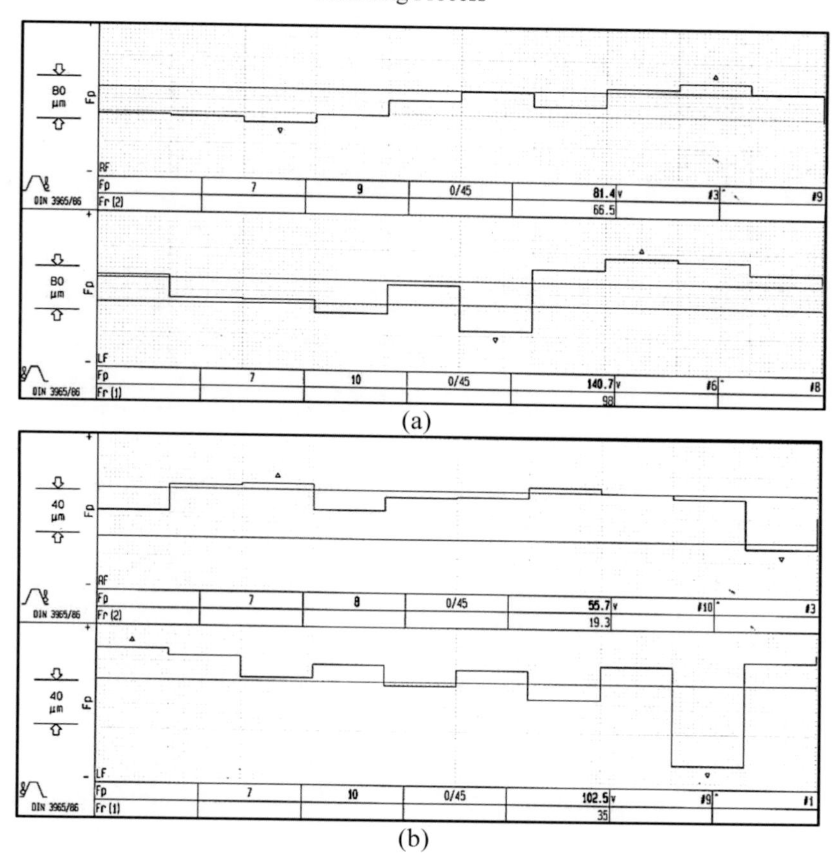

Total pitch error 'F_p' of the (a) unfinished straight bevel gear, and (b) Stage-3 best finished straight bevel gear

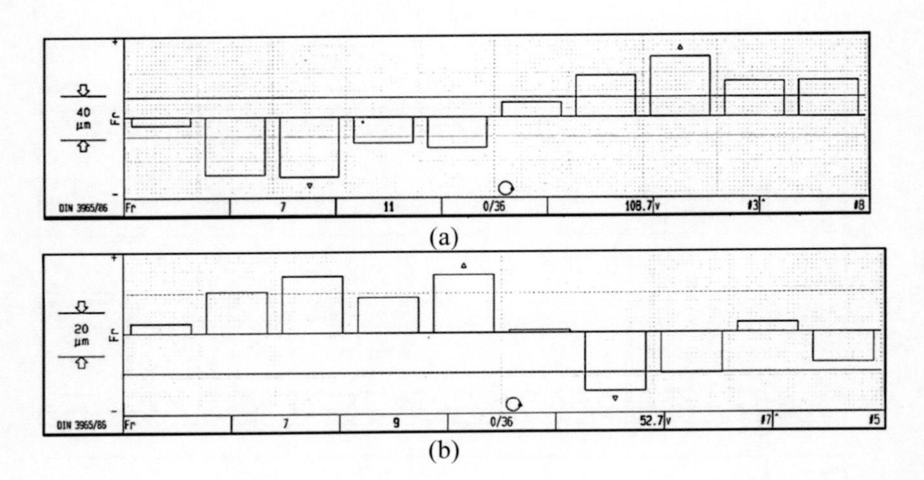

Radial runout 'F_r' of the (a) unfinished straight bevel gear, and (b) Stage-3 best finished straight bevel gear

(iii) Microgeometry graph of validation experiment

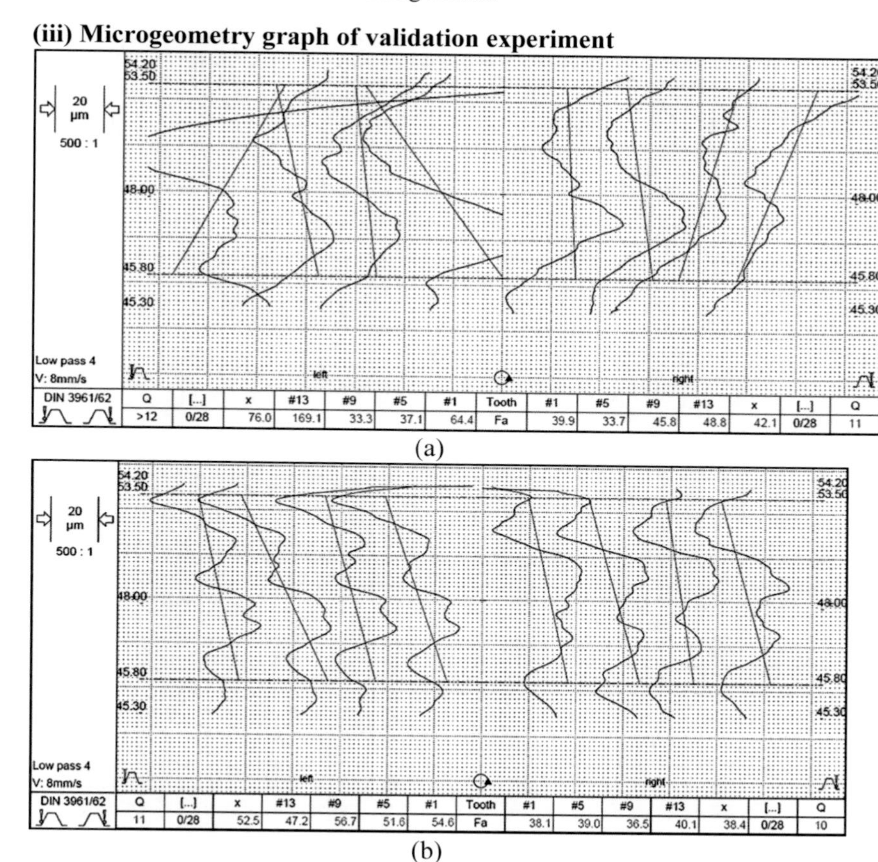

Total profile error 'F_a' of the (a) unfinished spur gear, and (b) AFF-finished spur gear using optimum process parameters given by DFA

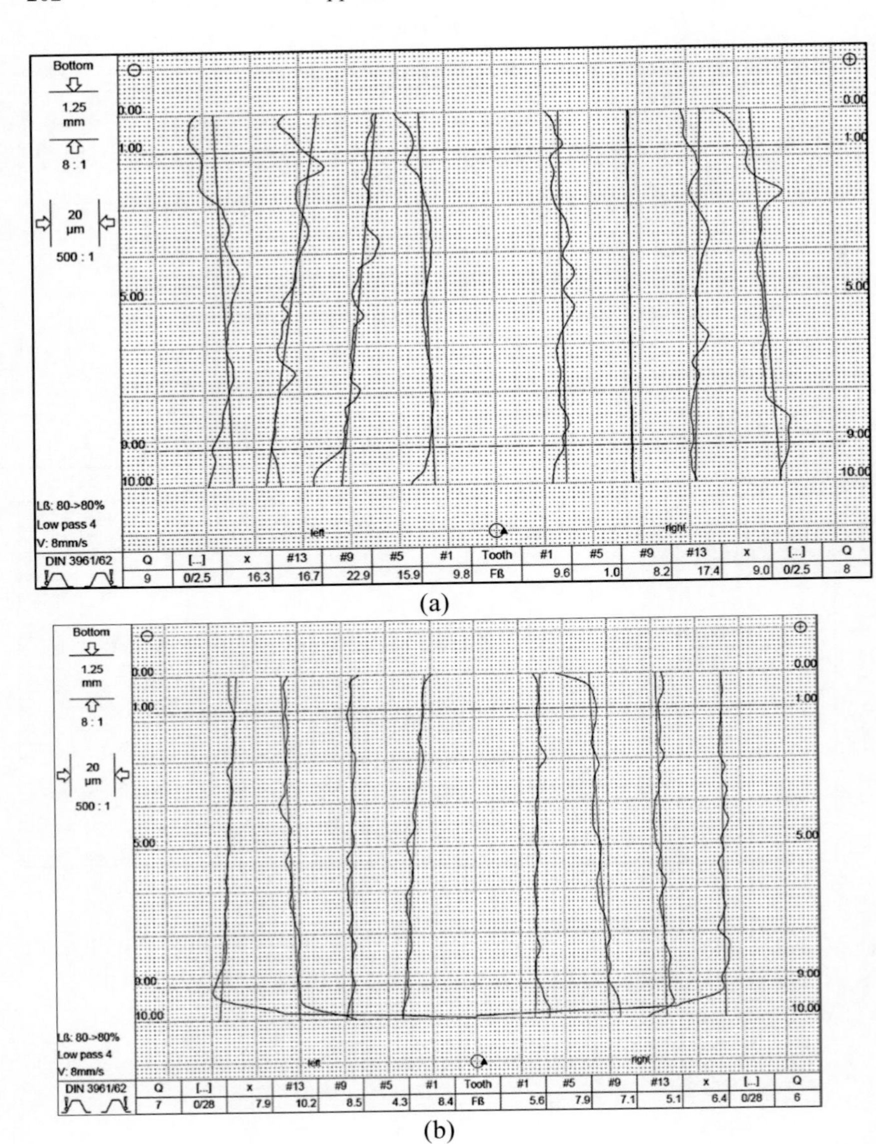

(a)

(b)

Total lead error 'F_β' of the (a) unfinished spur gear, and (b) AFF-finished spur gear using optimum process parameters given by DFA

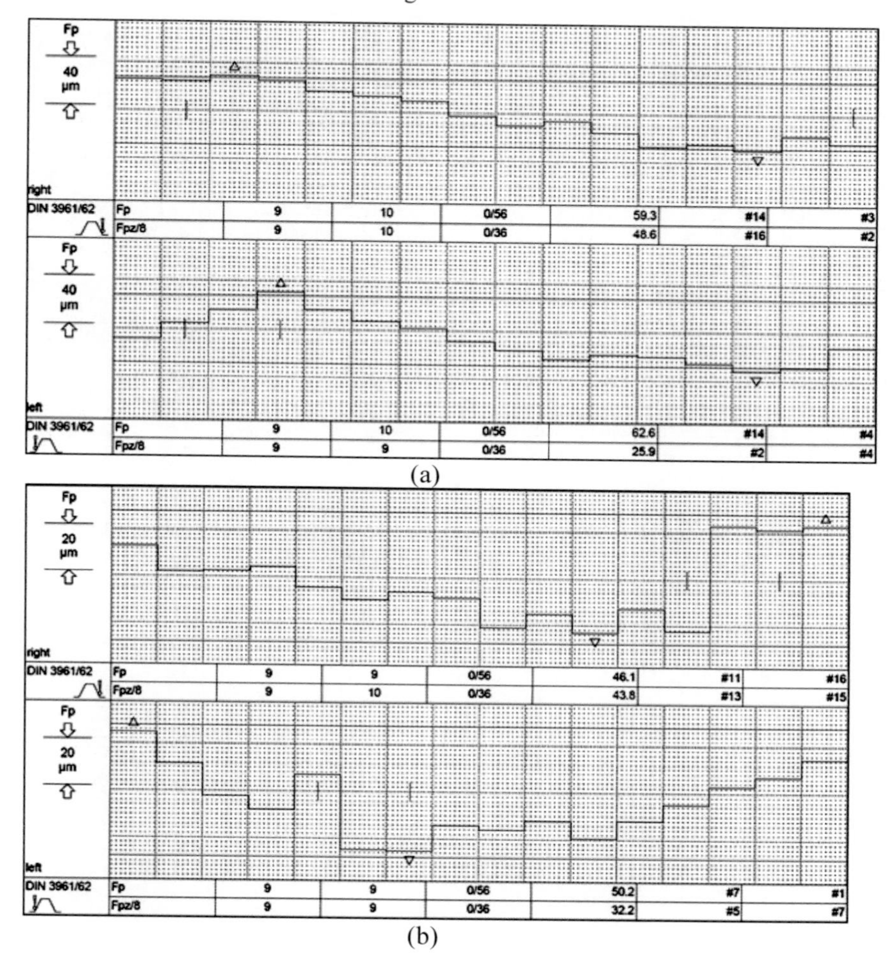

(a)

(b)

Total pitch error 'F_p' of the (a) unfinished spur gear, and (b) AFF-finished spur gear using optimum process parameters given by DFA

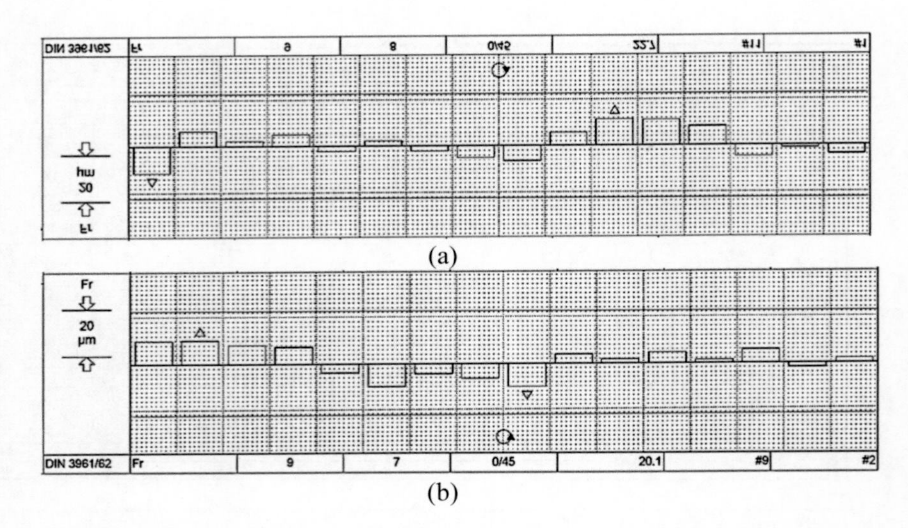

(a)

(b)

Radial runout 'F_r' of the (a) unfinished spur gear, and (b) AFF-finished spur gear using optimum process parameters given by DFA

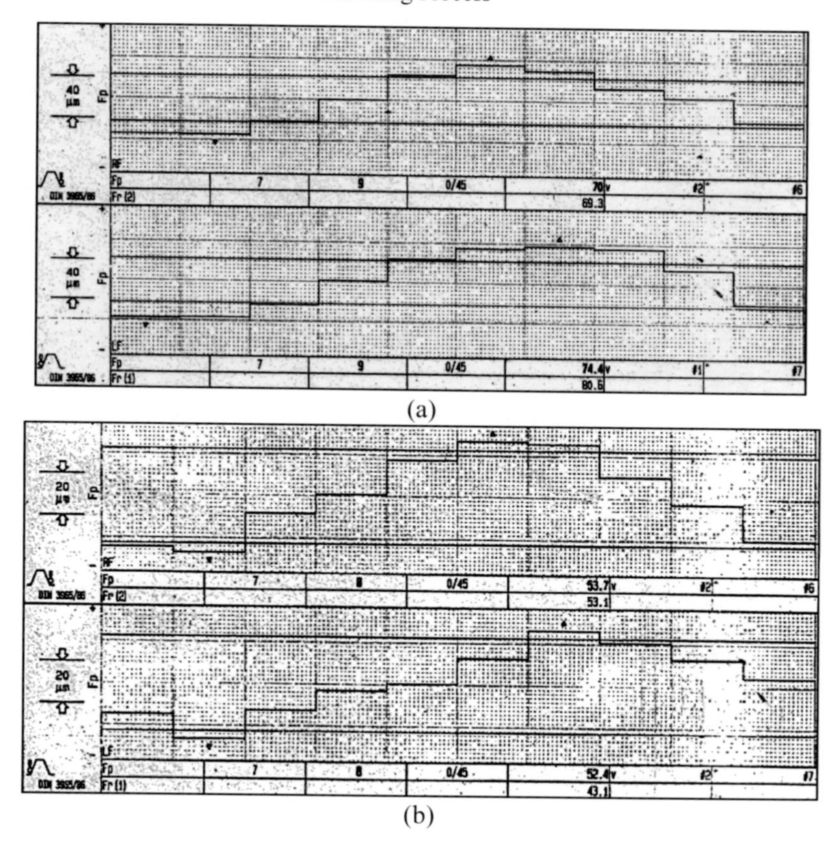

(a)

(b)

Total pitch error 'F_p' of the (a) unfinished straight bevel gear, and (b) AFF-finished straight bevel gear using optimum process parameters given by DFA

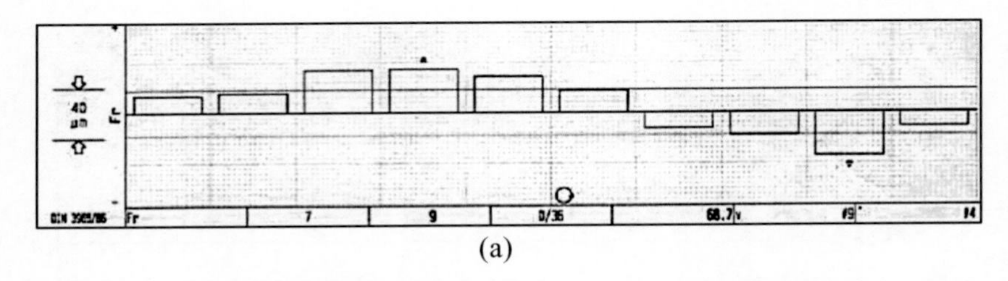

(a)

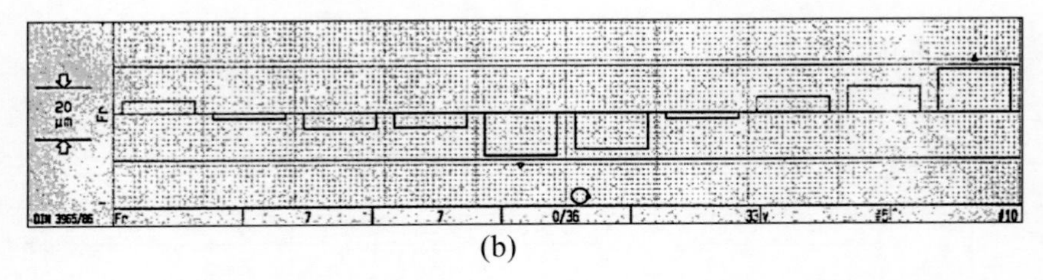

(b)

Radial runout 'F_r' for the (a) unfinished straight bevel gear, and (b) AFF-finished straight bevel gear using optimum process parameters given by DFA

(iv) Microgeometry graphs of AFF-finished the untextured and laser-textured spur and straight bevel gears

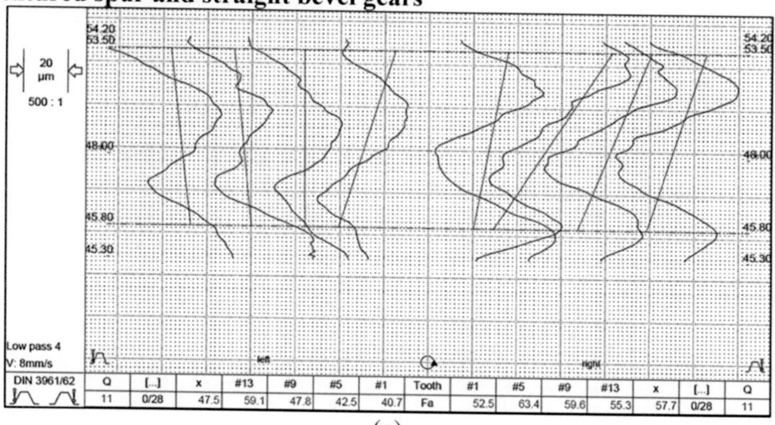

(a)

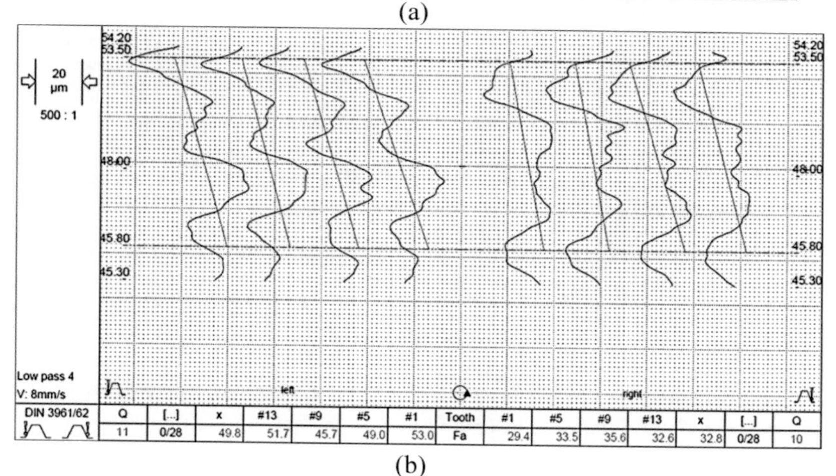

(b)

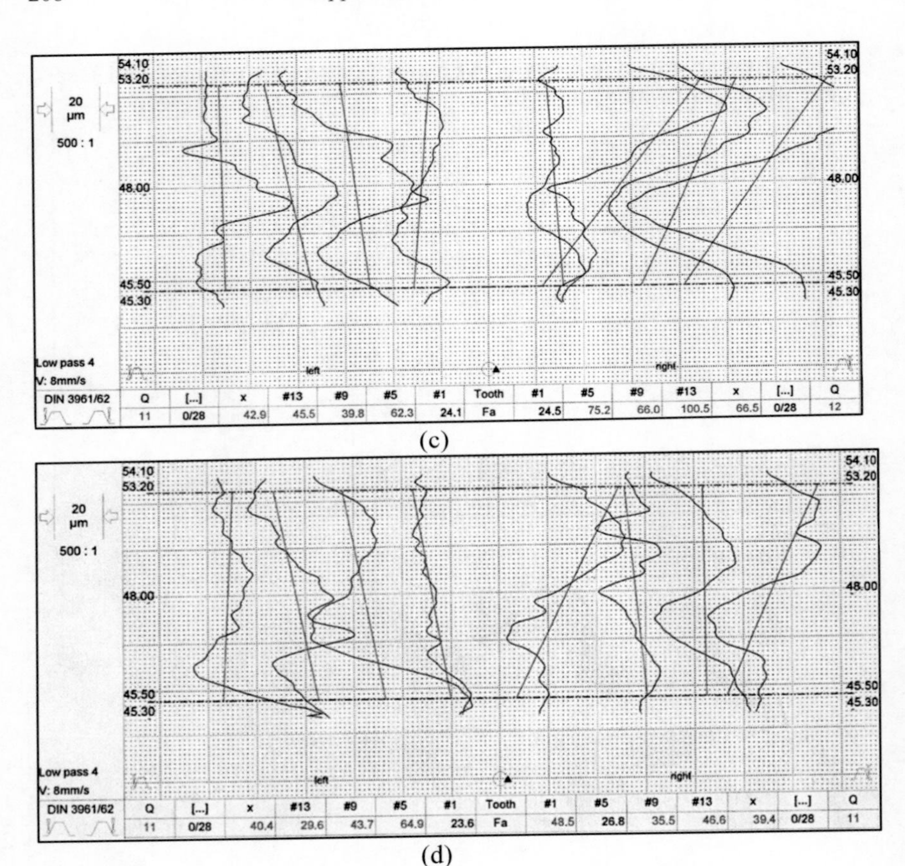

(c)

(d)

Total profile error *'F_a'* of the (a) untextured spur gear, (b) AFF-finished untextured spur gear, (c) laser-textured spur gear, and (d) AFF-finished laser-textured spur gear

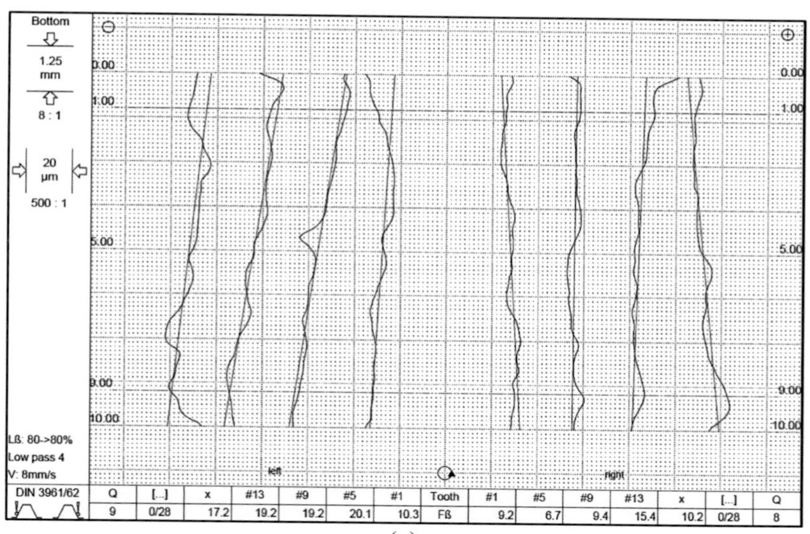
Bottom
1.25 mm
8 : 1
20 µm
500 : 1
0.00
1.00
5.00
9.00
10.00
Lß: 80->80%
Low pass 4
V: 8mm/s
left
right
DIN 3961/62
Q
[...]
x
#13
#9
#5
#1
Tooth
#1
#5
#9
#13
x
[...]
Q
9
0/28
17.2
19.2
19.2
20.1
10.3
Fß
9.2
6.7
9.4
15.4
10.2
0/28
8

(a)

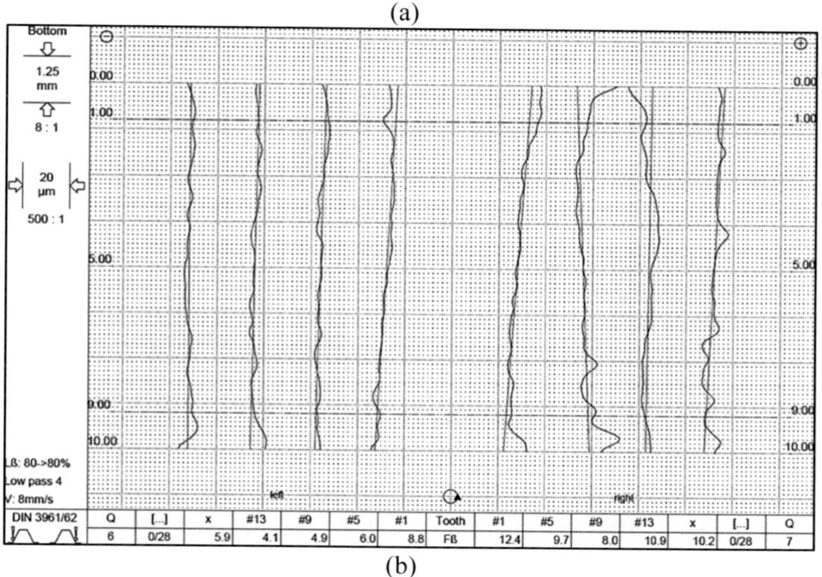
Bottom
1.25 mm
8 : 1
20 µm
500 : 1
0.00
1.00
5.00
9.00
10.00
Lß: 80->80%
Low pass 4
V: 8mm/s
left
right
DIN 3961/62
Q
[...]
x
#13
#9
#5
#1
Tooth
#1
#5
#9
#13
x
[...]
Q
6
0/28
5.9
4.1
4.9
6.0
8.8
Fß
12.4
9.7
8.0
10.9
10.2
0/28
7

(b)

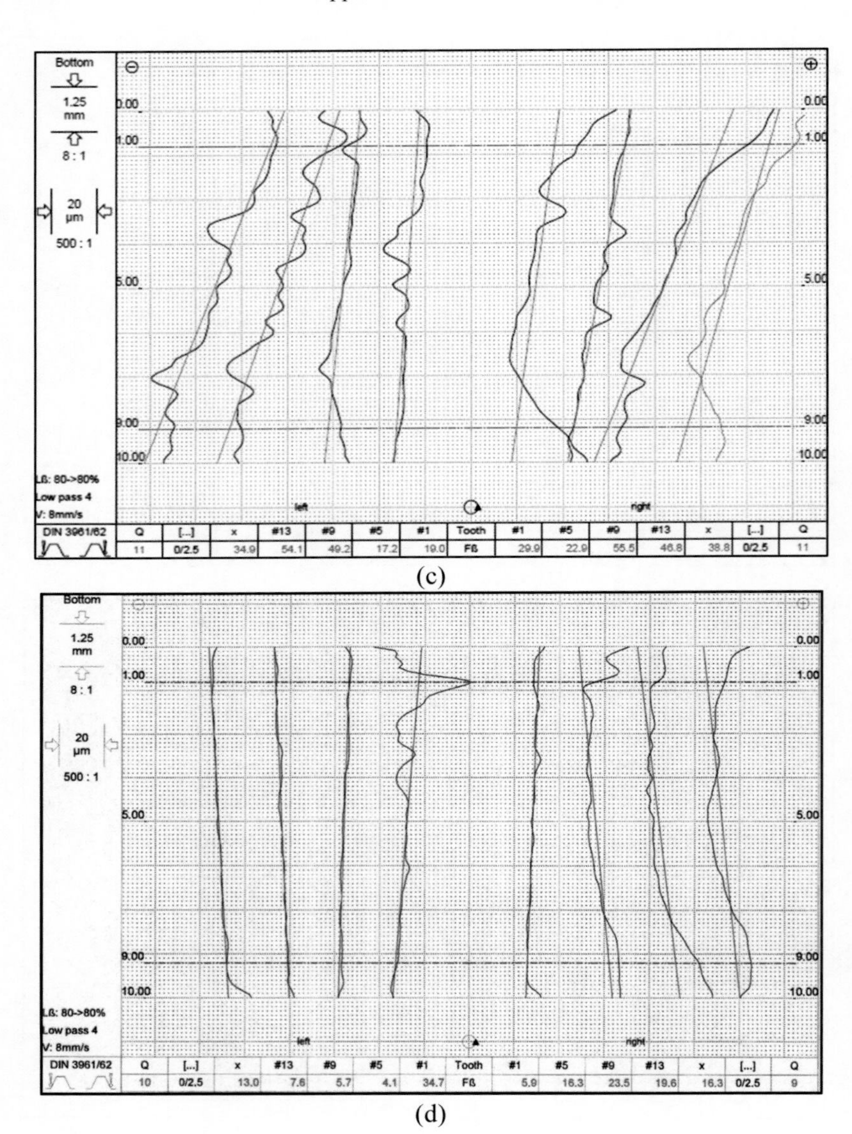

(c)

(d)

Total lead error 'F_β' of the (a) untextured spur gear, (b) AFF-finished untextured spur gear, (c) laser-textured spur gear, and (d) AFF-finished laser-textured spur gear

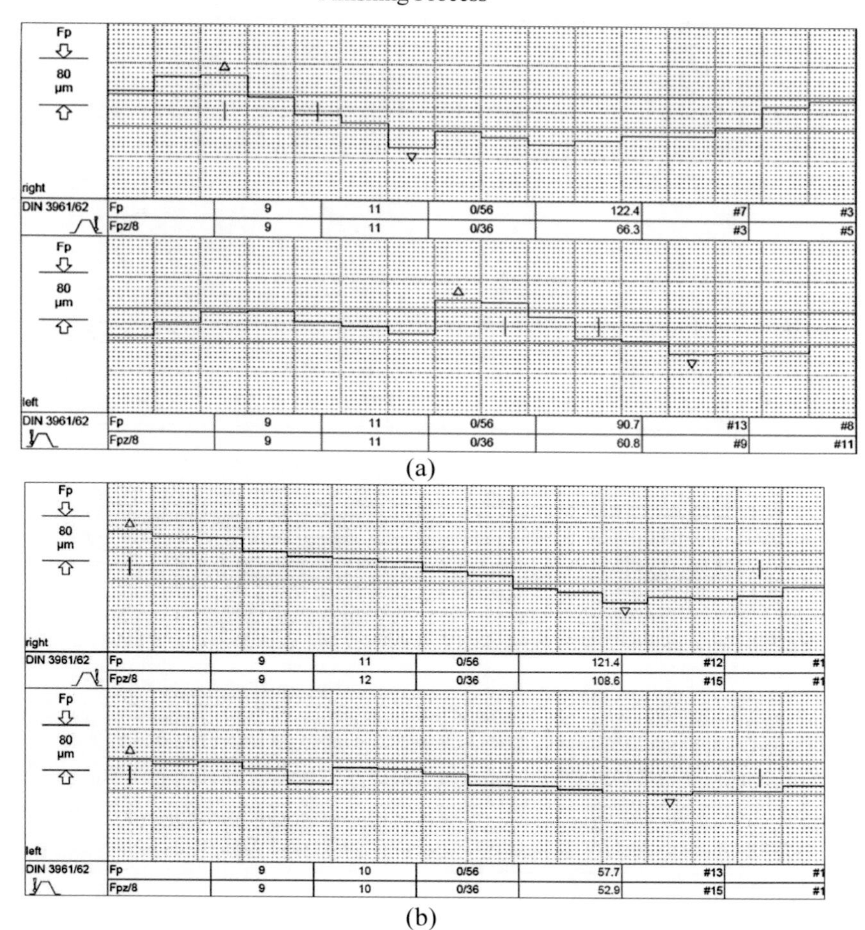
Fp
80
µm
right
DIN 3961/62
Fp
9
11
0/56
122.4
#7
#3
Fpz/8
9
11
0/36
66.3
#3
#5
Fp
80
µm
left
DIN 3961/62
Fp
9
11
0/56
90.7
#13
#8
Fpz/8
9
11
0/36
60.8
#9
#11
Fp
80
µm
right
DIN 3961/62
Fp
9
11
0/56
121.4
#12
Fpz/8
9
12
0/36
108.6
#15
Fp
80
µm
left
DIN 3961/62
Fp
9
10
0/56
57.7
#13
Fpz/8
9
10
0/36
52.9
#15

(a)

(b)

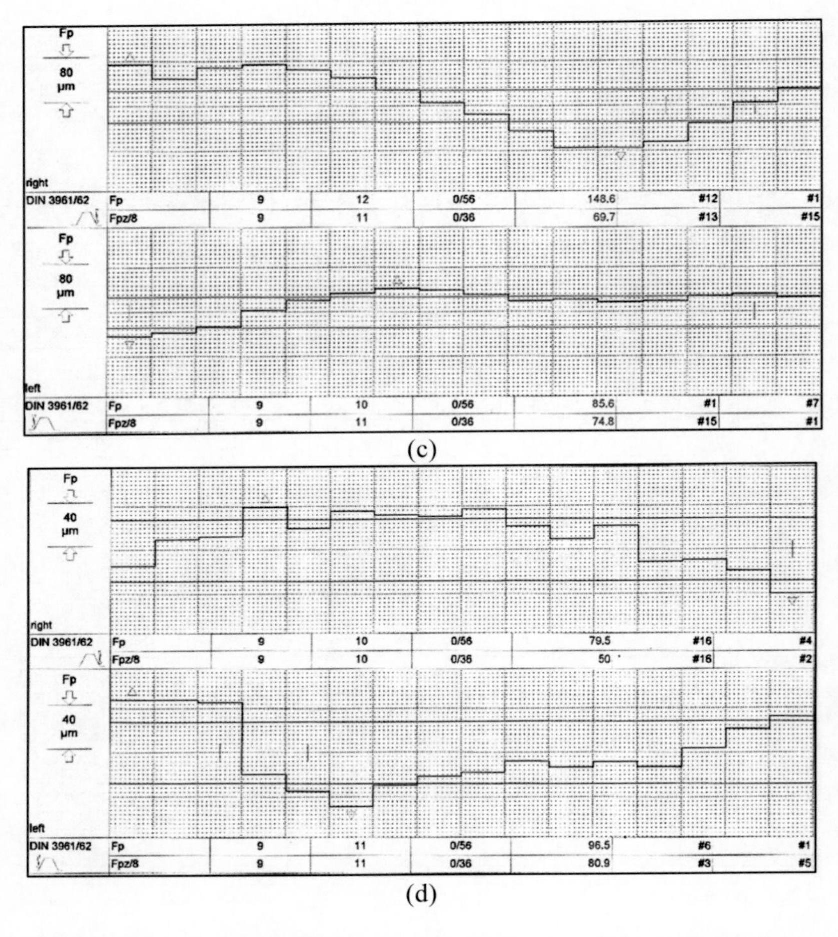

Total pitch error 'F_p' of the (a) untextured spur gear, (b) AFF-finished untextured spur gear, (c) laser-textured spur gear, and (d) AFF-finished laser-textured spur gear

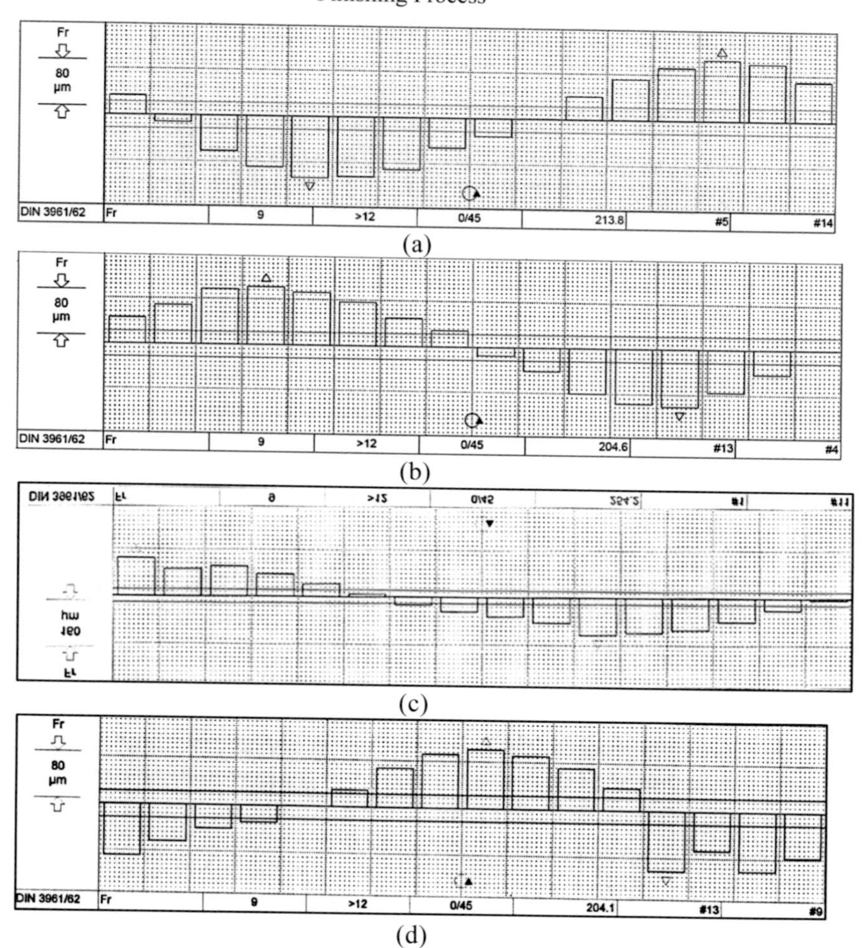

Radial runout 'F_r' of the (a) untextured spur gear, (b) AFF-finished untextured spur gear, (c) laser-textured spur gear, and (d) AFF-finished laser-textured spur gear

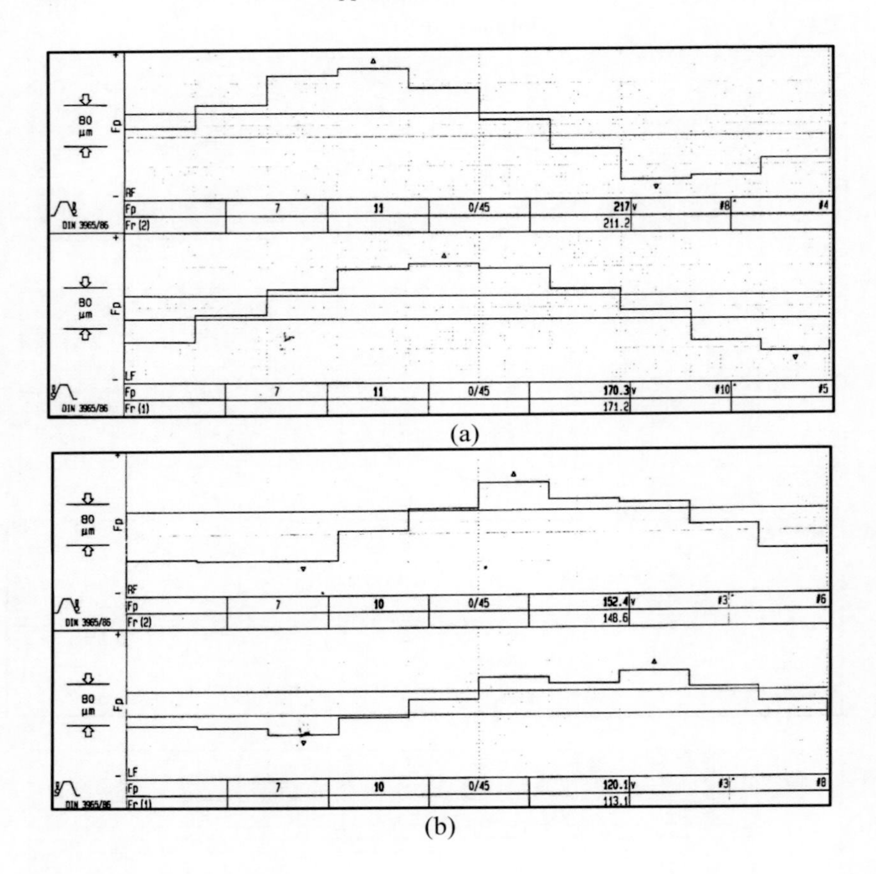
80
µm
Fp
RF
Fp
Fr (2)
7
11
0/45
217
211.2
#8
#4
DIN 3965/86
LF
Fr (1)
170.3
171.2
#10
#5
10
152.4
148.6
#3
#6
120.1
113.1
#8

(a)

(b)

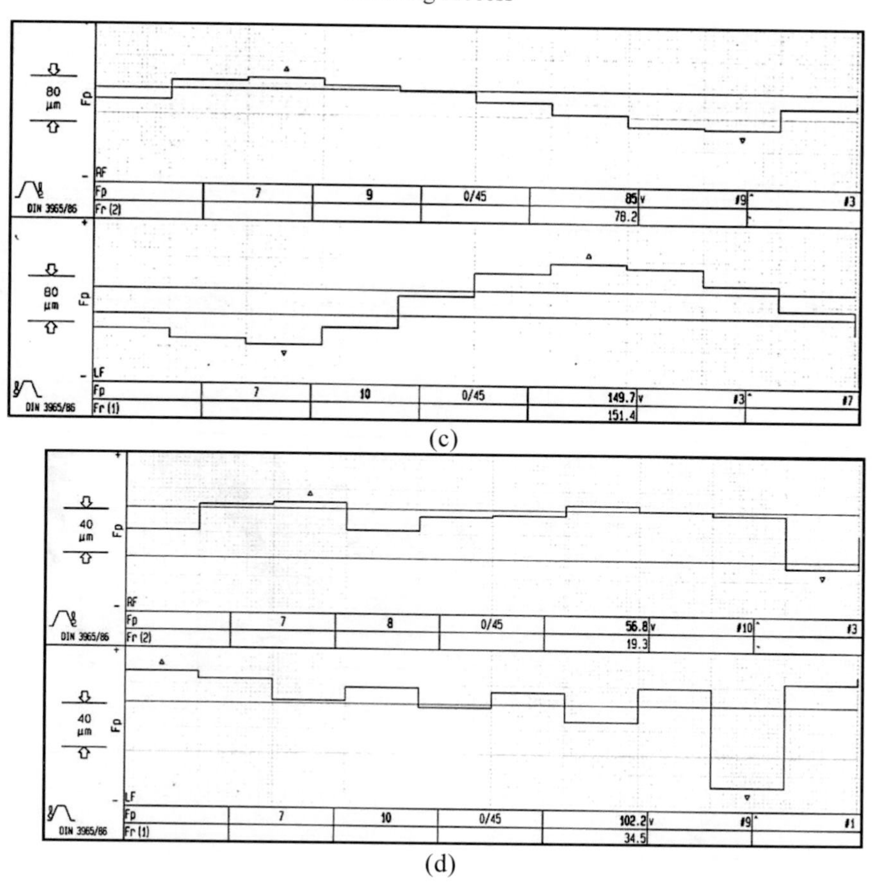

(c)

(d)

Total pitch error 'F_p' of the (a) untextured straight bevel gear, (b) AFF-finished untextured straight bevel gear, (c) laser-textured straight bevel gear, and (d) AFF-finished laser-textured straight bevel gear

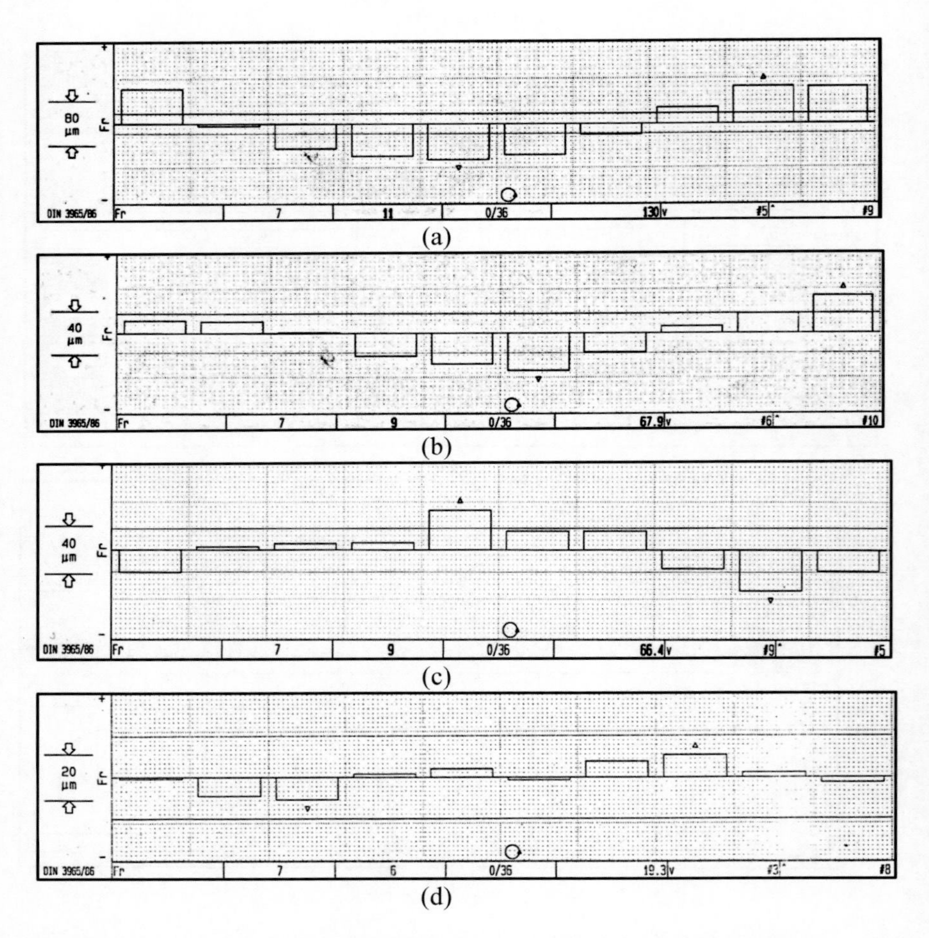

Radial runout 'F_r' of the (a) untextured straight bevel gear, (b) AFF-finished untextured straight bevel gear, (c) laser-textured straight bevel gear, and (d) AFF-finished laser-textured straight bevel gear

APPENDIX –D

DETAILS OF CONSTITUENTS OF THE AFF MEDIUM

(a) Abrasive particle

Type: Silicon Carbide (SiC)

Manufacturer: M/s Carborundum Universal Limited, Chennai (India)

Size: (i) **Stage-1 and Stage-2 experiments:** 100 mesh (with avg. diameter of 150 μm)

(ii) **Stage-3 experiments:** 80 (with avg. diameter of 190 μm); 100 (with avg. diameter of 150 μm); and 120 mesh (with avg. diameter of 127 μm)

(iii) **Validation experiments:** 120 mesh (with avg. diameter of 127 μm)

(iv) **Laser texturing experiments:** 100 mesh (with avg. diameter of 150 μm)

(b) Putty material

Type: Molding clay (silicon-based polymer)

Model and Make: Koolclay, Kores (India) Limited, Mumbai (India)

(c) Processing oil

Type: Silicone oil

Grade: Industrial grade

Density: 0.96 g/mL at 25 °C

Supplier: M/s S K Enterprise, Indore (India)

(d) Viscosity of AFF medium (kPa. S)

Before AFF (corresponding to vol. concentration of processing oil)	After AFF
135 (10%)	153
54 (15%)	61.5
8 (20%)	10.8
2 (25%)	5.8